PLATE IV. — Motor-driven Interrupter.

PLATE V. — The Electrolytic Rectifier.

PLATE II. — Transformer for 12″ Apparatus.

III. — Oscillation Transformer and Glass for Condenser of
12″ Coil.

THE TESLA HIGH FREQUENCY COIL

THE TESLA HIGH FREQUENCY COIL

ITS CONSTRUCTION AND USES

BY

GEORGE F. HALLER

AND

ELMER TILING CUNNINGHAM

56 ILLUSTRATIONS

NEW YORK

D. VAN NOSTRAND COMPANY

23 MURRAY AND   1910   27 WARREN STS.

i

The Plimpton Press Norwood Mass. U.S.A.

ii

PLATE I. — Complete 12″ Apparatus.   *Frontispiece.*

iv

INTRODUCTION

In presenting this book on the Tesla coil to the public the authors hope that they have filled a long felt vacancy in the practical library of science. No attempt has been made to give a mathematical explanation of the oscillation transformer, and other parts of the high-frequency apparatus, for the simple reason that the theory is too complex, and when obtained of no practical use. Neither have the authors tried to lead the amateur, who is just learning how to string bells and connect batteries, from the elements of the galvanic cell up to the working of a high-potential, alternating current, but have merely made an effort to place in the hands of advanced amateurs in electrical science a practical working manual on the construction of high-frequency coils, now so useful in scientific investigation.

The attention of the authors was first called to the Tesla coil when they were fortunate enough to be given the use of the 7″ standard coil described in the last chapter of this book. A systematic line of experiments was carried on with it, in order to study the effects of a change in the constants of the various circuits. All the mechanical and electrical details of construction were carefully worked out, and the authors finally decided to design and construct a larger coil. The coil, as first constructed, was a decided failure, due to

too small a condenser capacity.   For about five months they
further experimented on the details of construction and
finally arrived at the 12″ coil described in this book.   This
coil they feel assured is as efficient as can be made.   It is
especially designed to give a high-frequency discharge of
great volume.   This latter fact makes it useful for wireless
telegraphy.

In conclusion they have to thank Mr. G. O. Mitchell for
many suggestions and for the kindly interest he has taken in
this work.   They feel that without his help the writing of
this little book would have been impossible.

G. F. H.

E. T. C.

CONTENTS

CHAPTER   PAGE

    I. GENERAL SURVEY 1

   II. THE TRANSFORMER 4

  III. THE CONDENSER 20

  IV. THE OSCILLATION TRANSFORMER 24

   V. THE INTERRUPTER 32

  VI. THE CONSTRUCTION OF THE BOXES 60

 VII. ASSEMBLING 64

VIII. THEORY OF THE COIL 72

  IX. USES OF THE COIL 84

   X. DIMENSIONS OF 7″ STANDARD COIL 97

APPENDIX 111

x

LIST OF FIGURES

FIG.   PAGE
1.   Method of Fastening Primary Terminals — Completed Primary   8
2.   Secondary Bobbin of Transformer   11
3.   Hand Winder   13
4.   Wire-spool Holder   14
5.   Frame for Secondary of Transformer   18
6.   Section of Completed Transformer   18
7.   Condenser Frame and Brass Condenser Sheet   22
8.   End Support for Secondary of Oscillation Transformer . . .   25
9.   Fibre Strip   25
10. Centre Rod   25
11. End Support for Primary   28
12. Primary of Oscillation Transformer   29
13. Completed Secondary of Oscillation Transformer   29
14. Bushings for Support of Oscillator Standards   30
15. Hard Rubber Block on Oscillation Transformer   30
16. Simple Primary Air-gap   34
17. Magnetic Interrupter   37
18. Motor Interrupter Fan   38
19. Brass Angle Piece   39
20. Hard Rubber Block   40
21. Section of the Motor Interrupter   41
22. Patterns of Base   42
23. Patterns of Yoke   44
24. Section of Completed Motor   45
25. Rotor Disc   46
26. Rotor and Clamp Nut   48
27. Stator Disc   49
28. Frame for Stator Coils   51
29. Self-starting Device   52
30. Rectifier Plates and Wiring Diagram   58
31. Transformer Box   61
32. High-tension Box   63
33. Connections for Primary of Transformer   65

FIG.		PAGE
34.	High-tension Bushing	66
35.	Oscillators and Standards	67
36.	Wiring Diagram	68
37.	Waves on Wires	92
38.	Primary and Core of Transformer of 7″ Coil	98
39.	Secondary Bobbin of Transformer of 7″ Coil	99
40.	Plate and Frame of Condenser	101
41.	Oscillation Transformer of 7″ Apparatus	104
42.	Box for 7″ Apparatus	106
43.	Wiring Diagram	107
44.	Oscillators and Standards for 7″ Apparatus	109
45.	Oscillation Transformer of Small Coil	113
46.	Completed Transformer of Small Coil	115
47.	Primary Spark-gap	117
48.	Wiring Diagram	118
49.	Wiring Diagram	118

THE TESLA COIL

CHAPTER I

GENERAL SURVEY

By far the largest and most interesting branch of science is electricity, for Maxwell has proven mathematically, and Hertz verified experimentally, that light is an electromagnetic disturbance in the ether, and thus added that subject to the realm of electricity. Amongst the various phenomena of electricity, those of the high-tension current are the most interesting and instructive. With such a current all the wonders of the Geissler and Crookes tubes may be seen. With it waves for wireless messages may be sent out into space, and a great number of other experiments carried out.

It is the purpose of this book to show how a satisfactory apparatus for producing these currents may be constructed, and also to describe a few of the uses for such a coil.

The apparatus, as described in this book, is most commonly known as the Tesla High-Frequency Coil, and consists, in general, of four parts: 1. The Step-Up Transformer; 2. The Interrupter; 3. The Condenser; 4. The Oscillation Transformer. Each of these will be fully considered in subsequent chapters.

Before entering upon the description of the Tesla high-frequency apparatus, however, it would be well to make a few general remarks which are of the greatest importance.

Throughout the whole work of construction the most exacting care must be given to the matter of insulation. All the wire used must be carefully tested, and each layer of wire in the transformer must be thoroughly shellacked, and then insulated from the next layer, by two turns of carefully oiled paper. In the condenser, which is really the vital part of the apparatus, the glass should be of the best grade obtainable. It must also be free from all air bubbles. It is in the high-frequency apparatus, however, that the greatest care as regards both construction and insulation must be taken. The secondary consists of one layer only of No. 32 B. & S. gauge, double cotton-covered wire, wound on an octagonal frame, formed of strips of vulcanized fibre fastened to two end pieces of wood. When winding the wire, care must be taken that no two adjacent wires touch, for that would cause a short circuit. When the wire is completely wound, it is given about five coats of shellac, not only to act as an insulator, but also to prevent any slipping of the wires. The primary consists of a thin band of copper, making two and a half turns around a circular frame surrounding the secondary. The frames on which the primary and secondary are wound must be very firm and substantial, so that an occasional jar will not displace any of the wires on the secondary.

All connections must be soldered, and the connecting wires run through glass tubes.

When the apparatus is finished, two carefully made boxes must be constructed. These must be oil tight. This is accomplished by mortising all joints, and then giving the boxes, especially the joints, about four or five coats of shellac. Into one box the transformer fits, and into the other the condenser and oscillation transformer. Then the boxes are filled with pure paraffine oil, which is the only efficient insulator for these high-tension currents.

Some who intend to build this coil will think that all these precautions regarding insulation are extreme, but it will be found that, in dealing with high-frequency, high potential currents, too much care cannot be taken, for "Good insulation is the key to success in high tension work."

CHAPTER II

THE TRANSFORMER

THE transformer — sometimes called a converter — is merely an induction coil that is connected directly to the alternating-current mains, without the use of an interrupter, and is used to raise or lower the voltage. In a transformer the number of watts in the primary equals approximately the number of watts in the secondary.

In the case of any step-up transformer, the ratio of the number of volts in the primary to those set up in the secondary is nearly the same as the number of turns of wire in the primary to the number in the secondary; but the amperes decrease in the inverse ratio.

The transformer used in the coil described in this book is of the common induction-coil type, oil-immersed, step-up transformer. It takes the alternating current from the mains at 110 volts or 55 volts, and steps it up to about 10,000 volts.

The efficient working of a transformer depends largely upon the design of the core. The iron used must be of high permeability and should have little retentivity. A straight core is always best to use; for, on the fall of the current from its maximum value to zero, the magnetic flux falls from its maximum value, not to zero, but to a value which depends

4

on the residual magnetism. The residual magnetism in an open circuit is much less than in a closed magnetic circuit, so that when the current suddenly becomes zero, the magnetic flux drops lower in an open circuit than in a closed one. As the electromotive force in the secondary is proportional to the fall in the magnetic field, it is greater with a straight core than with a closed circuit of iron.

The coil designer is obliged to determine the length of the iron core from the experience of others, as the mathematics for calculating it is too complex, although simple and useful in the case of closed circuit transformers. If the core is made too long the primary magnetizing current will be too large, while if made too short the secondary coils would have to be made of too large a diameter to be efficient. There is, therefore, a certain length which will give the best results.

In the case of this transformer the length of the core was determined after having gained all possible information from certain eminent men who had made a life study of these matters; in fact, all the dimensions of the transformer for this special use were determined in this way.

The iron core is made up of pieces of No. 20 or 22 B. & S. gauge iron wire 18″ long. The wire is first cut nearly to size with a pair of pliers, and, when assembled, the ends of the bundle are sawed off square with a hack saw. An ordinary piece of iron pipe, a little less than 18″ long, and having an internal diameter slightly less than 2″, is tightly filled with these wires. When putting the wires in, stand the pipe on end on a smooth surface, and force in each wire until it hits this surface. When the bundle is finished, the upper end is sawed off with a hack saw to exactly 18."

The tube containing the iron wires is now placed in a coke or coal fire and left there until the fire burns itself out, thus insuring slow cooling. This heating and subsequent slow cooling so softens the iron wires that their retentivity is reduced to a minimum. When cool, the wires are taken out and sandpapered to remove any superfluous oxide. They are then, one by one, dipped into boiling water, wiped dry, and while still warm are coated with thin shellac varnish. When the shellac is dry they are again packed, as tightly as possible, in the pipe, to hold them in the desired shape. Then, while still packed closely together, they are forced slowly out of the pipe; starting at the end thus released, they are tightly bound with a narrow cotton bandage, which can be obtained from any surgical supply house. The bandage should be between one and two inches wide, but no more. When the entire core is wrapped with this cloth, the cloth should be heavily shellacked. The ends of the core are now filed flat and smooth; after this it is put in a warm place to dry thoroughly, when it will be ready for the primary winding. The use of the insulating varnish on the iron wires is to arrest eddy currents as much as possible, thus preventing the iron wire from becoming heated and energy wasted, which would lower the transformers efficiently.

The primary is wound in two sections of two layers each, one above the other. No. 12 B. & S. gauge, double cotton-covered copper wire is used. About $2\frac{1}{4}$ pounds will be required. The primary may be wound by hand, by erecting two wooden supports 17" apart, and having a 2" hole bored in each, to receive the iron core. Then, by turning the core by hand the wire may be wound fairly well. But as it is rather difficult to wind the wire tightly in this way, it would be more satisfactory to wind it in a lathe, if the amateur has access to one. To mount it, cut a half-inch piece from the end of the pipe in which the core was formed, and slip it over the extreme end of the core. Make the ring fit as tightly as possible by placing between it and the core a few strips of tin or other thin sheet-metal. Now clamp it firmly in the chuck. The other end of the core should also be fitted with a half-inch piece of pipe and supported at this place in the steady rest. The one piece of pipe is used to prevent any of the wires from being forced in unequally at the points where the chuck clamps it, and the other to afford a smooth bearing surface for the steady rest. If there is any tendency for the core to slip out of the chuck, the tail stock, with the centre removed, may be pressed up against it.

About 1 ft. from the end of the copper wire take a couple of turns of tape around it. At this point bind the wire to the iron core, about 1" from its end, by taking several turns of tape around it. Proceed now to wind the wire tightly and closely to within 1" of the other end. Here the winding of the primary is stopped for a short time in order to give the wire a good coat of shellac. After the shellac has dried, another coating is given it, and then the second layer is wound on while the wire is still wet. When the winding

has come to within about six turns of the starting point, a
piece of tape doubled back on itself is laid on the first layer,
with its ends projecting beyond the unwound portion of the
second layer. The looped end of the tape must be on the
outer side of the winding. See Fig. 1.

 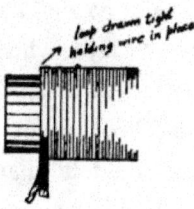

Diagrams showing manner of fastening last turn.

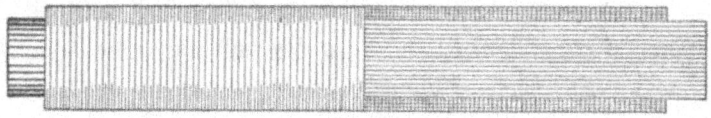

FIG 1. — METHOD OF FASTENING PRIMARY TERMINALS — COMPLETED
PRIMARY.

The winding of the second layer is finished over the piece
of tape, the last turn being brought through the loop in the
tape. The loop is drawn tight by pulling on the other projecting ends.   In this way the last turn is kept from slipping
off. By using this method or fastening it is unnecessary to
use any bobbin heads for the primary; this is a decided
advantage, as, with a removable primary, bobbins are always
getting loose. The wire is cut off about 2′ from this ending

in order to allow plenty of wire for making the various connections, which will be described in a later chapter. When this layer is thoroughly shellacked, the first section of the primary is complete.

The second section is wound directly on top of the first, starting at the same end, and being sure to wind in the same direction. Each layer when wound is thoroughly shellacked, and the last turn is fastened in the same manner as before. If the wire has been put on carefully 164 turns can be wound in the 16″ and the total diameter will be $2\frac{5}{8}$″.

The secondary is wound in four sections. It will first be necessary to procure two micanite tubes, the one fitting tightly within the other. The inner diameter of the smaller tube is a trifle greater than $2\frac{5}{8}$″, the external diameter of the larger one being $3\frac{1}{4}$″. The length of the tubes is 18″ and their thickness $\frac{1}{8}$″. Now turn out a wooden rod so that the larger tube will fit around it tightly. Mount the rod in the lathe with the tube on it, clamping one end of the wood in the chuck, and supporting the other end on a centre. With a thin parting tool, cut off seven rings, three 1″ wide, and four, $3\frac{1}{4}$″ wide. If the amateur has no lathe the rings may be cut off in a mitre box. Out of some quarter-inch sheet-fibre, cut eight circular pieces, 6″ in diameter and having a 3″ hole in the centre. Slip one of the 1″ rings on the smaller tube, and with Le Page's glue fasten it to the extreme end of the tube. Next slip on one of the circular discs of fibre, and then one of the $3\frac{1}{4}$″ rings, fastening them with glue. Two more discs are put on, and then another $3\frac{1}{4}$″ ring. After this comes another disc and a 1″ ring, followed by a disc and a 3¼″ ring. Then put on two more discs and the remaining 3¼″ ring. This is followed by the remaining disc and 1″ ring. Be sure that each ring is carefully glued in place. Before putting on the discs, small holes should be drilled in them, through which to carry the wires. The completed bobbin for the secondary is seen in Fig. 2. The discs numbered 2, 3, 6, 7 have the holes for the connecting wires drilled on their inner edge, while the others have them drilled about ½″ from their outer edge. Obtain a wooden rod upon which the secondary bobbin will fit tightly. It should be 18½″ long.

If the coil builder is skilled in winding wire in the lathe, the winding may be done there much more rapidly than by hand; but for an amateur, who has had but little experience with lathe winding, or for one who does not possess a lathe, the following method is given. In winding in the lathe, great care must be taken that the wire is not snapped off when the end of the layer is reached, and while the paper is being wrapped on before the next layer is wound.

For the hand winder, the wooden rod, on which the secondary bobbin fits tightly, is drilled in at both ends for about 4″ with a little less than a ¼″ hole. Pieces of ¼″ iron are then driven into these holes, to serve as an axle. They should fit tightly, so as to turn with the cylinder. About 6″ should project at one end, which is bent into a handle. 1½″ at the other end is sufficient for a bearing.

The standards are made of ¾″ oak, fastened 19¼″ apart,

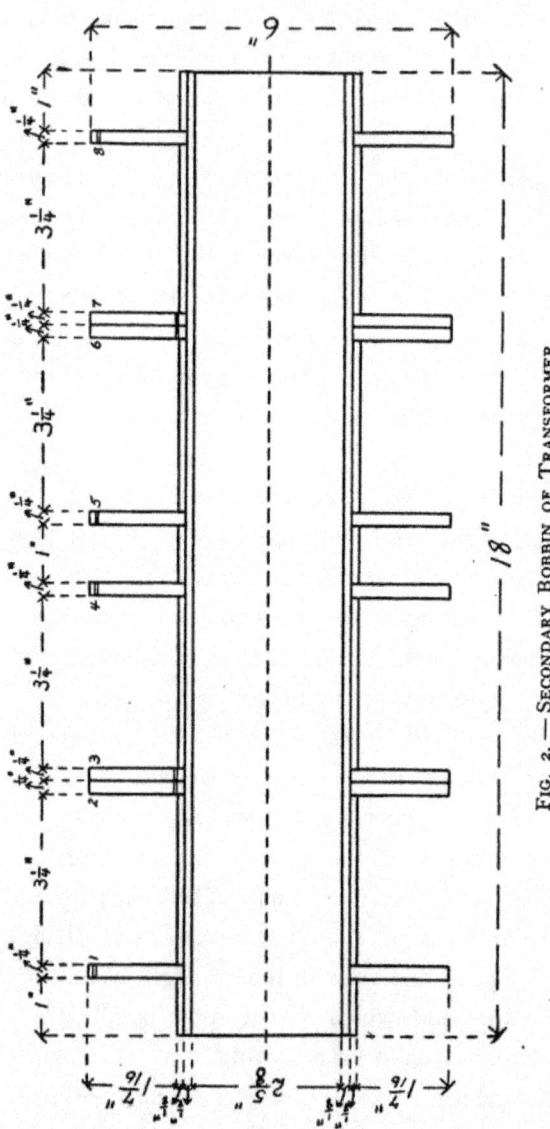

FIG. 2.—SECONDARY BOBBIN OF TRANSFORMER.

to a baseboard 2′ long. A piece of oak $\frac{3}{4}$″ square and 2″ long is fastened with two screws to the top of each standard, to serve as a cap. A $\frac{1}{4}$″ hole is then bored with its centre on the joint. This allows the cylinder to be taken out of its bearings when necessary. Two iron washers are slipped over the shaft at the short end to act as a thrust bearing, and two washers, with an open, steel-wire spring between them, are put on the other end. This will give the friction required to enable the amateur to stop the winding at any time, and still be sure that the cylinder will not rotate and so loosen the turns of wire. The dimensions of the winder are seen in Fig. 3.

As the wire must be wound under some tension, and as it is tiresome to give the required tension by letting the wire run through the hand, the holder shown in Fig. 4 was devised.

It consists of an axle which fits the spool tightly, and which is 4″ longer than the spool. There is a thread cut on one end of this axle for about 2″. It is then mounted in two wooden standards fastened to a baseboard. Iron washers are put between the spool and the standards for the spool to bear on. An open spring made of piano wire is slipped up on the threaded end of the shaft, outside of the standards. A washer and a nut are now put on to give the required tension to the spring. A lock nut is put on to keep this nut from turning.

Care must be taken to detect any breaks that may occur in the wire. When winding the wire it quite frequently

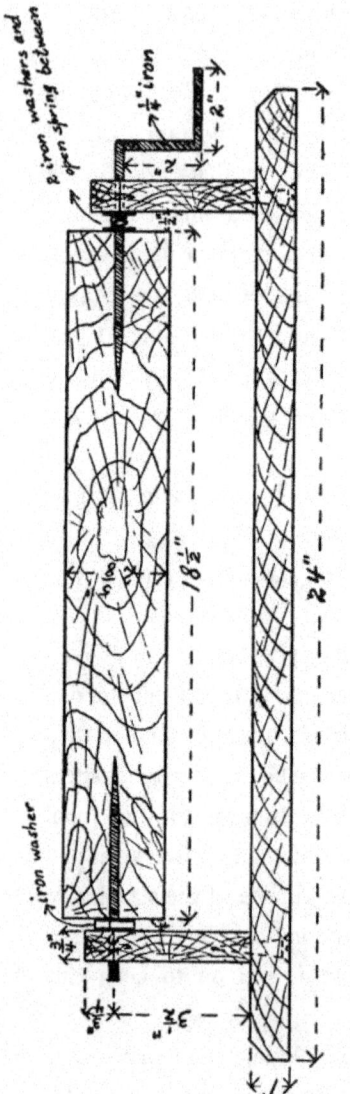

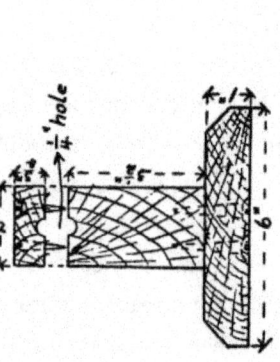

Diagram showing hand winder for winding secondary.

Note – Not drawn to scale.

FIG. 3. — HAND WINDER.

happens that a little kink will cause a break; but because
it is covered by the cotton insulation, it will be wound on the
bobbin, unknown to the coil builder. To detect these breaks
immediately, the authors used the following method. A ring
is cut out of a piece of sheet brass or copper. It is ¼″ wide
and 3″ in diameter. This is fastened by several flat-headed
brass screws to one side of the spool on which the wire is

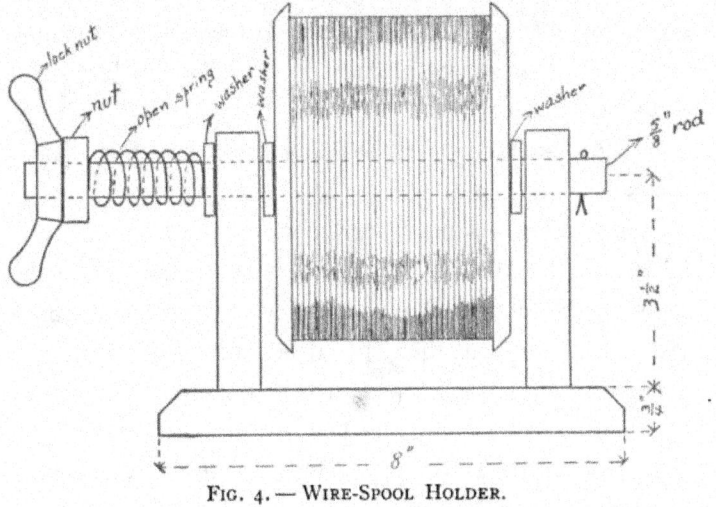

FIG. 4. — WIRE-SPOOL HOLDER.

bought. If the wire has been bought from a reliable dealer,
the inner end will be found projecting outside of the reel.
This wire is soldered to the ring on the outside of the spool.
A strip of sheet copper or brass, which is of such a length
that it will bear on the ring, is fastened to the upper end of
the standard, on the side on which the ring is. From here
a wire is led to one pole of a dry cell.

On the winder a strip of sheet metal is fastened to one of the standards. It is best to fasten it to the one farthest up from the handle. It is bent so that it presses firmly on the projecting axle, which has been polished to make good electrical contact. A wire is then led from the brush on the standard to a binding post on the baseboard. A telephone receiver is now connected in series with the binding-post and the other pole of the cell. A watch-case receiver, with a head attachment, is the best to use. If the amateur has only the Bell receiver, an attachment to hold it to his head can easily be arranged.

If the amateur prefers he may use a sensitive galvanometer.

Everything is now ready for the winding of the secondary. To begin, pass about 1′ of the wire through the hole in the bobbin heads numbered 2, 3, from the side on which bobbin head 2 is. The insulation should be scraped off of the end of the wire for about 2″, and then this bare part should be tightly wrapped on the axle between the washer and the wooden cylinder. It will now be seen that there is a complete circuit through all the wire on the spool. The diaphragm in the telephone receiver is drawn down or if a galvanometer is used, the needle will be deflected. If the wire should break, the diaphragm will return to its normal position and a click will be heard, or in the case of the galvanometer the needle will return to the zero position. When this happens the break should be located and the wire soldered. Acid should not be used in soldering, as a little left on the wire will corrode it and spoil the electrical connection. Rosin is the best thing to use as a flux.

The first layer in the section between the bobbin heads 1, 2, is wound from 2 to 1, and after it is wound it is given a good coating of shellac. Before winding the next layer, a little over a turn of paper is taken around the previous one. The edge of the paper can be held down with a little shellac. Paraffine wax must not be used to increase the insulation, as the transformer when finished is immersed in paraffine oil, which acts as a partial solvent to paraffine wax, thus spoling its insulating properties. All the layers after the first should start about $\frac{1}{4}''$ from the inner face of the discs and stop the same distance from them. Be sure to shellac each layer after it is wound and then take a turn of paper around it. Continue winding until 61 layers are in place. The last layer should be wrapped over with a narrow cotton bandage which is thoroughly shellacked to keep it in place About two feet of wire should be left projecting from the section for making the various connections. This wire is then brought through the hole in disc 1, and its end is connected to the axle.

Unwind the other wire from the axle and, after polishing the uninsulated part with a piece of emery cloth, twist it around the end of the wire from the spool, which has also been polished, and then solder the connection. Wrap the bare part of the wire with some silk thread, so as to thoroughly insulate it.

The section between the bobbin heads 3, 4, is now wound.

The first layer is wound from 3 towards 4, the winder being turned in such a direction that the direction of the current in the wire of this section will be the same as in the one just wound. That means that the winder must be rotated in the opposite direction. For convenience, however, in the winding the secondary bobbin can be taken off of the wooden rod and put back in a reversed position. When this is done the winder is rotated the same as previously. The same instructions hold for this section as for the previous one. Each layer must be shellacked and wrapped with paper and the winding must stop before getting to the bobbin heads. In this section 71 layers are wound on. About 2′ of wire should be brought through the hole in disc 4 to allow for connections.

The two remaining sections are now wound in the same manner, there being 71 layers in the section between the bobbin heads 5, 6, and only 61 layers in the section between the discs 7, 8. Remember to keep the direction of the wire the same as in the previous coils. This method of winding has several advantages, one of them being that it relieves static strains. A practical reason is that all the leading out wires are from the outer layers, thus making it always easy to bring out a new piece of wire if any are ever broken off.

The reason for having a greater number of layers in the middle coils will readily be seen from a consideration of the direction and intensity of the lines of magnetic force around a solenoid.

Leave the completed secondary in a warm place to thor-

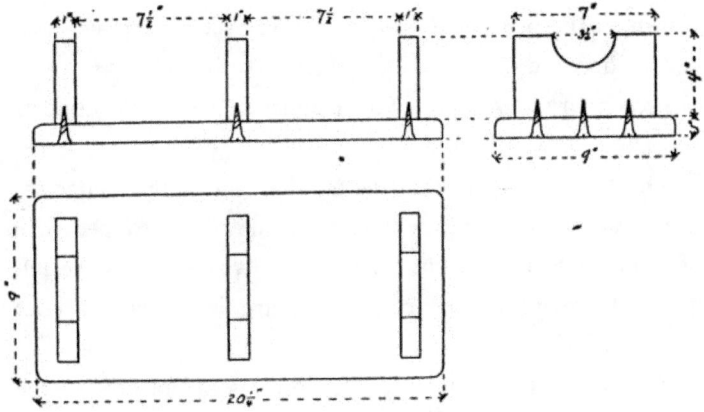

FIG. 5. — FRAME FOR SECONDARY OF TRANSFORMER.

oughly dry, and in the meantime construct the frame for the secondary. Working drawings are given in Fig. 5. The base is made of a piece of 1″x 9″ pine 20¼″ long. The corners and edges are rounded to give it a better appearance. Three supports of 1″x 7″ pine, 4½″ high, are erected at the points

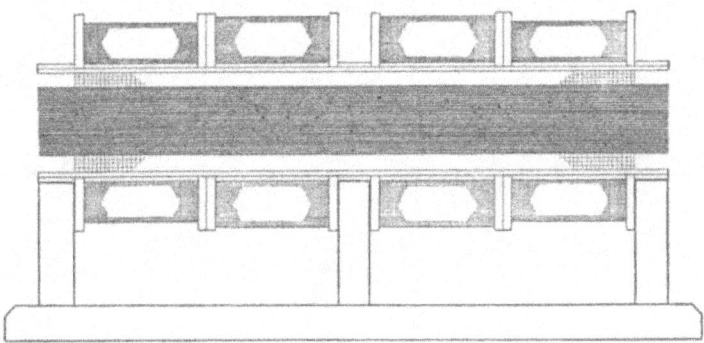

FIG. 6. — SECTION OF COMPLETED TRANSFORMER.

shown in the figure. At the top of each a half-round hole $3\frac{1}{4}''$ in diameter is cut. Into these fits the tube on which the secondary is wound. The distance between the supports is $7\frac{1}{2}''$.

The two terminals from the middle coils of the secondary are now soldered together, and the connection wrapped with silk thread to insulate it. The other two terminals are left alone at present as their connections are described in a later chapter.

The reason for using a cotton cloth instead of tape in the construction of the transformer is that oil almost immediately spoils all the adhesive qualities of the type.

PLATE II. — Transformer for 12″ Apparatus.

III. — Oscillation Transformer and Glass for Condenser of
12″ Coil.

CHAPTER III

THE CONDENSER

A CONDENSER is an apparatus for accumulating a large quantity of electricity on a small surface. The form may vary considerably, but in all cases it consists essentially of two conductors separated by a non-conductor or dielectric, and its action depends entirely upon induction.

The thinner the dielectric and the greater its specific inductive capacity, the greater is the capacity of the condenser. A thin dielectric, however, cannot withstand a high potential. Besides the thickness, the dielectric strength depends on the character of the material.

The condenser used in this apparatus is especially designed for long continued use on high voltages. The dielectric used is glass and the plates are made of sheet brass. When finished the condenser is immersed in pure paraffine oil.

For the dielectric 95 sheets of glass, $\frac{1}{10}''$ thick and $10'' \times 12''$ in size, should be obtained. These sheets may be had cut to size for about nine or ten cents apiece. In purchasing them each sheet should be examined carefully to see if there are any air bubbles. If any are found in a sheet of glass it should be rejected.

The brass used is number 32 or 34. Forty-six sheets

8″ x 10″ are required. In one of the corners of the shorter side is a tongue 2″ wide and 1¼″ long. A ¼″ lip is bent across the top of this tongue. If the brass can be had in rolls 8″ wide, a little more should be obtained and the tongues cut out of it. They should be cut 3″ x 1¼″. The extra 1″ is for soldering them to the plates. Rosin, not acid, should be used as a flux. These tongues should always be soldered in the position shown in the figure.

As a rule 12″ is the only width that can be obtained in most places. When this width is used the tongues are cut right on the sheets.

The condenser occupies the part that has been constructed for it in the oscillation transformer box. The frame in which the condenser is built up is made out of well dried pine. The base is made of a piece of ½″ x 11″ pine, 11¾″ long. The sides are cut out of ½″ x 12″ material 10¾″ long. The ends are also cut from ½″ x 12″ wood and are 11″ long. The sides and ends should be planed up smooth on both sides, so as to make them a little less than ½″ thick. The completed frame is seen in Fig. 7.

Place the condenser frame on a table or some other flat surface, with one of the ends down. Before putting any of the glass sheets in the frame, they should be carefully wiped clean so as to remove any dust or moisture. Commence by putting two glass sheets in the frame so that they reach the bottom of the frame. Place a brass sheet on top of these glass plates so that there is a 1″ margin of glass all around the sheet, except were the tongue comes out. If the lip on the tongue has been bent carefully it will just fit up against the sheets of glass. Without displacing the brass lay two sheets of glass on top of it. A brass sheet is next put in, but in this case the tongue comes out on the reverse side. There should be a 1″ margin around the brass as in the previous case. After the brass come two more sheets of glass. This process is

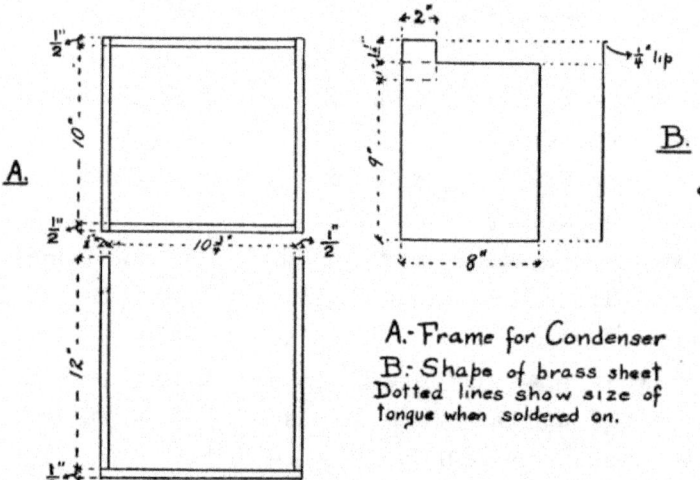

A.-Frame for Condenser
B: Shape of brass sheet
Dotted lines show size of
tongue when soldered on.

FIG. 7. — CONDENSER FRAME AND BRASS CONDENSER SHEET.

kept up until the 46 sheets of brass have been put in place. Three sheets of glass are placed on top of the last brass plate. If the glass and the brass used are the size called for in this book the last sheet will just go in tightly. In forcing in the last few sheets it is a good idea to lay a cloth between the last two, to take up any excess pressure, which would otherwise crack the glass.

Set the condenser upright, when finished, and solder a piece of No. 16 bare copper wire about 3' long to each of the lips in turn, down the one side, and the same is done on the other side.

Two leather straps should be fastened to the sides of the frame to lift it by when lowering into the oscillation transformer box. The object for building the condenser in a separate frame, instead of in the division in the oscillation transformer box, is to facilitate moving should the condenser ever require rebuilding, due to the rupturing of the glass sheets.

When the condenser is placed in the box, the end which has the three glass sheets should be placed against the partition, that is, nearest the oscillation transformer. This is to prevent the spark from the oscillation transformer breaking through into the condenser, instead of following its air path.

CHAPTER IV

THE OSCILLATION TRANSFORMER

It is this part of the apparatus, so simple in construction, in which the most care as regards insulation must be taken. The success of the whole apparatus depends on the care with which this part is constructed. The least fault, such as two wires touching, or many other small similar mistakes, may cause a short circuit and require its reconstruction. It is not wise to hurry the work, as it will be necessary to reconstruct it if careless.

The end supports are made out of any suitable piece of wood. The two supports for the secondary are 8″ in diameter, and from ¾″ to 1″ in thickness. Eight equidistant points are marked off on the periphery and slots ½″ deep and ½″ wide cut at these points. See Fig. 8. These slots are for the fibre strips, on which the secondary is wound, to fit into. These strips are 17″ long and ½″ square and are cut from the best vulcanized fibre obtainable, eight being required. In each end of the strips a hole is drilled and countersunk to receive a small brass screw, which is to fasten them to the end pieces. A wooden rod about 1″ in diameter is now obtained and a shoulder turned on each end. The diameter of the shoulder is ½″ and its length just equals the width of

the supports, *i.e.*, ¾″ if ¾″ wood is used, or 1″ if 1″ wood is
used for the ends. The length of the rod over all must be

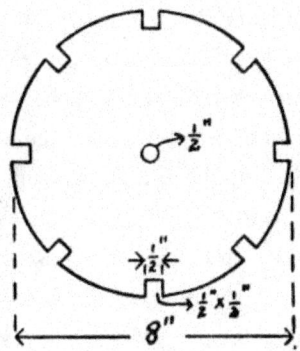

FIG. 8. — END SUPPORT FOR
SECONDARY OF OSCILLATION
TRANSFORMER.

only 17″. See Fig. 10. A ½″ hole is now drilled in the centre
of each of the end pieces. These pieces are now slipped on,

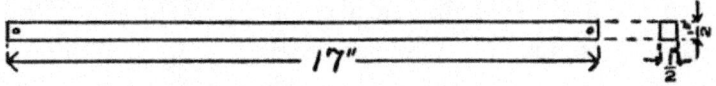

FIG. 9. — FIBRE STRIP.

one on each end of the rod. Screw the fibre strips on the
supports, after fitting them in the slots and seeing that they
are parallel to the rod in the centre. If any of the strips are

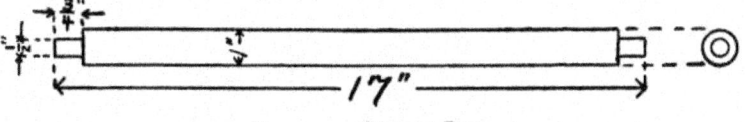

FIG. 10. — CENTRE ROD.

bent or warped they should be straightened before being fastened in place. A good way to straighten them is to lay them between two boards, placing some heavy weight on the top board, and leave them thus over night.

About 4 or 5 ounces of No. 28 B. & S. gauge double cotton covered copper wire is required for the secondary. There are several equally good ways in which the wire can be wound on the frame. Two of these methods will be described, as the authors have found them both satisfactory.

The first method is intended for those that have a lathe at their disposal. A cylinder of wood 4″ in diameter and 18″ long is first turned out, and on it are screwed the eight strips of fibre, so that their ends are in line, and that they strike one another. If the amateur has a couple of clamps made of strips of sheet iron with a bolt through the ends, it will greatly help matters by clamping them around the strips, about 6″ apart, and moving them as needed. A light cut is first taken off the strips, and then they are polished with a file and sandpaper. A No. 18 thread is now cut, starting about 1″ from the end of the strips to within an inch of the other end. It should be cut just as deep as possible. In order to make a clean cut the tool must be very sharp and several light cuts should be taken instead of one. When the strips are mounted on the frame again it will be seen that there is a continuous groove in which the wires will lie without touching one another.

The secondary frame is now supported between centres in the lathe so that it just turns easily. Around one end of one of the strips wrap about 1′ of No. 28 wire. This is to be used for connections. Starting at this point wind the wire tightly on the frame, always keeping the wire in the groove cut for it. About 1′ extra should be left at the end for making the connections. When finished, a heavy coat of shellac is given to the wires where they rest on the fibre strips. When this is dry all the wire is heavily shellacked. A soft brush should be used so as not to displace any of the wires. The secondary should now be placed in a warm place to dry.

The following method can be used in case a lathe is not available. A spool of silk thread and some silk are required. There must be enough silk to make two turns around the secondary frame. It is wrapped on tightly, and smoothly, and shellacked in place. It might be mentioned here that shellac dissolved in wood alcohol should not be used. The wire is then wound on, starting about 1″ from the end up to within an inch of the other end. A silk thread is wound on at the same time between the turns, to keep them apart. When it is all wound the wire is heavily shellacked and the frame put in a warm place to thoroughly dry.

Get two pieces of $\frac{3}{4}″ \times 12″$ pine 12″ long and find the centre of each by the intersection of their diagonals. With these points as centres describe two circles, with radii $4\frac{1}{2}″$ and $5\frac{1}{2}″$ on each board. Divide the circles having the radius $5\frac{1}{2}″$ into 36 equal parts. These points can be located with a protractor or mariner's compass, there being one every 10°.

At each of these points drill a ¼″ hole. In each corner of these boards cut out a piece 1″ x 1″. See Fig. 11.

At a planing mill buy eighteen ¼″ dowels, 36″ long. These may be had for one cent apiece. Each one should be cut in two, which will give you thirty-six ¼″ rods, 18″ long. Fasten one of the 12″ end pieces to each of the ends of the

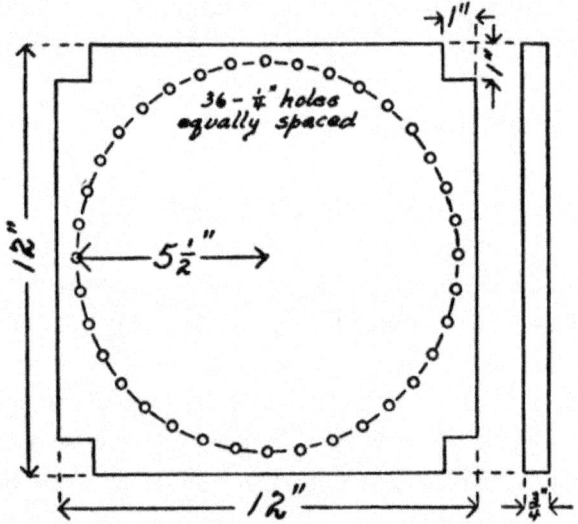

FIG. 11. — END SUPPORT FOR PRIMARY.

completed secondary frame, by four flat-headed brass screws. See that the circular end pieces of the secondary just fit on the smaller circles drawn on the boards. The ¼″ rods are now driven into place, thus forming a circular cage on which the primary is wound.

For the primary get an 8′ piece of No. 36 copper ribbon ½″ wide. Wrap the end of the ribbon once around the end of the dowel shown in Fig. 12 and solder it in place. A piece of copper wire should also be soldered on here. Starting from here take two and a half turns around

FIG. 12. — PRIMARY OF OSCIL-
LATION TRANSFORMER.

the secondary frame. This will bring you to the end of the dowel diametrically opposite the dowel from which you started. Wrap the ribbon once around this dowel and solder it in place. A piece of wire should also be

FIG. 13. — COMPLETED SEC-
ONDARY OF OSCILLATION
TRANSFORMER.

soldered on. The turns of the ribbon should be equally spaced. Get four pieces of 1″x1″ pine 18½″ long and fasten them as braces from the one end of the frame to the other, the pieces fitting in the corners that were cut out for them.

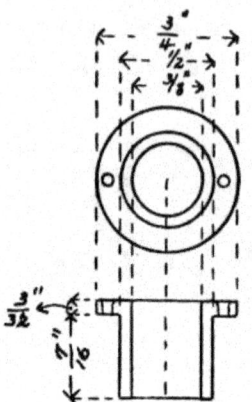

FIG. 14. — BUSHINGS FOR
SUPPORT OF OSCILLATOR
STANDARDS.

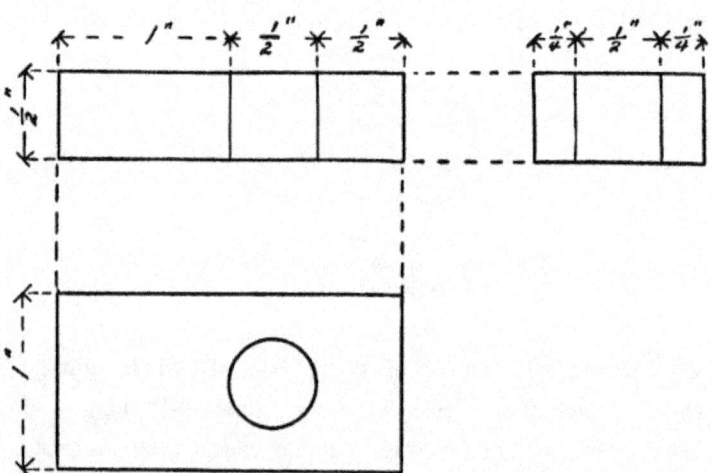

FIG. 15. — HARD RUBBER BLOCK ON OSCILLATION TRANSFORMER.

Turn two bushings out of a piece of $\frac{3}{4}''$ brass or copper rod. They are shown in Fig. 14. The length over all is $\frac{1}{2}''$ and the shoulder is $\frac{1}{8}''$ thick. A $\frac{3}{8}''$ hole is drilled down the centre, and the shoulder is drilled and slightly countersunk at two opposite points to receive two small brass screws. Two pieces of vulcanized fibre or ebonite $\frac{1}{2}'' \times 1''$, $2''$ long, with a $\frac{1}{4}''$ hole drilled $\frac{1}{2}''$ from the end, are fastened to the middle of the top side of the primary frame. The bushings just made are fitted into the holes and screwed in place. The wires from the secondary are soldered onto these bushings. When this is done the oscillation transformer is finished and all that needs to be done is to connect it up properly. The completed oscillation transformer is seen in plate III.

CHAPTER V

THE INTERRUPTER

LET us now consider that important part of the apparatus, the primary spark-gap. The function of this, as previously stated, is to provide a path of high resistance until the condenser is charged to its full capacity. Then it suddenly breaks down, and allows the current to surge back and forth across it, until the current is damped out by resistance and other factors in the circuit. After the oscillations have ceased, the ideal spark-gap should return to its maximum disruptive strength before the condenser can be charged by the next cycle from the secondary of the transformer.

In practice this is far from being the case. The air between the discharging balls becomes heated, and offers a comparatively low-resistance path for the current. This results in an arc being formed, which prevents the condenser from performing its function. The mechanical problem which confronts the designer is to find some way to get rid of this heated air and thus prevent the arc being formed. The object of this chapter is to show several ways of partially accomplishing this result, but in no case perfectly.

The simplest form of spark-gap for the primary of the oscillation transformer, one which has given fair results with the writers, consists of merely two adjustable brass balls. No provision is made for blowing out the arc that forms, so that considerable of the energy of the transformer is wasted. Nevertheless sufficient oscillations are set up to bring the coil to within 1″ or 2″ of its maximum length of spark discharge.

Most amateur coil builders will, at this point in the construction, be very anxious to see what the possibilities of their work are. A good idea of the working value of the apparatus may be obtained with the simple air-gap. When a designer is more acquainted with the workings of a Tesla coil he can construct that one of the spark wipe-outs which is best suited to the current that he has at his disposal.

THE SIMPLE AIR-GAP

Two standards of brass $\frac{3}{4}$″ in diameter and 4″ long are mounted on a piece of hard rubber. They should be about 5″ apart. A $\frac{3}{16}$″ hole is drilled $\frac{1}{2}$″ from the upper end of each rod and is taped with a standard $\frac{1}{4}$″ machine screw tap. These two holes must be in line.

Next two pieces of $\frac{1}{4}$″ brass rod $3\frac{3}{4}$″ in length are threaded for the whole of their length to fit the holes tapped in the standards. Two balls $\frac{1}{2}$″ in diameter are turned out of brass or are procured from a dealer in physical supplies. These balls are drilled and threaded to fit in the brass rods. Two vulcanized fiber rods are turned out of $\frac{3}{4}$″ rod, 3″ in length, and a $\frac{1}{4}$″ shield 2″ in diameter is screwed on one end of each handle. The handle is drilled and tapped for

1" to fit the brass rod on which it is screwed. The shield is to safeguard the operator's hand from sliding off the fibre handle and coming into contact with the transformer current, which would probably be fatal. Connection can be made to the brass standard near the bottom by drilling

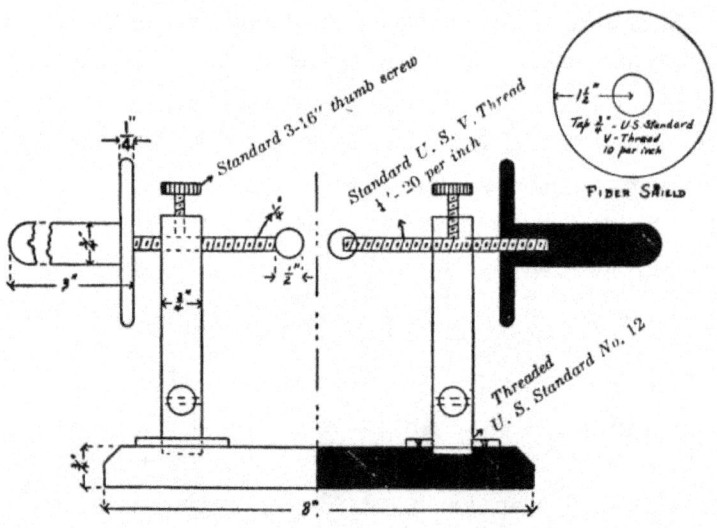

FIG. 16. — SIMPLE PRIMARY AIR-GAP.

a small hole through them, and then drilling and tapping another hole at right angles to the first, for a thumb screw to bind the wire.

A set screw at the top of the standards to clamp the rods in place after they have been adjusted will be a convenience to the operator. See Fig. 16.

THE AIR-BLAST INTERRUPTER

In order to make the air-gap more efficient, getting rid of much of the heated gases between the spark terminals, a mechanical means can be used of forcing in cold air, thus driving out the heated gases and keeping the resistance much higher. To efficiently accomplish this a piece of $\frac{3}{8}''$ hard glass tubing is drawn out into a nozzle having an opening $\frac{3}{16}''$ in diameter. This is mounted on a brass standard and is connected by means of a rubber tube to a foot bellows such as is used in the laboratory to operate a blast lamp or any other suitable supply of compressed air. A good blast of air will effectually wipe out any arc that tends to form, thereby increasing the disruptive length of the bright oscillation transformer discharge. The operator will find, however, that it is a tedious task to pump a foot bellows, occupying so much of his time as to handicap him in performing experiments with the high-frequency discharge, and he will soon decide that the best policy is to construct either a magnetic wipe-out or a motor-driven interrupter.

THE MAGNETIC INTERRUPTER

The magnetic blow-out is well suited for those who have a source of direct current at their disposal; either the 110-volt lighting circuit or a suitable battery current. For those who have only the alternating current and who wish to use the magnetic wipe-out the writers have added to this chapter a simple home-made electrolytic rectifier of their own design which will give a current suitable for magnetizing the magnet of this interrupter.

Two standards of the same form as those used in the simple interrupter, but 6″ in length and having spark balls ¼″ instead of ½″ in diameter, are mounted on a hard rubber base 10″ x 7″. The fibre handles and shields are also necessary for this interrupter. Two electromagnet bobbins 5½″ long and having an iron core ¾″ in diameter with fibre heads 3″ in diameter are procured. An iron yoke made from 1″ x ¼″ iron 7″ long has 2 holes drilled in it in the middle 1½″ from both ends, and the bobbins are fastened to it by a screw in the core. Two polar pieces of ¾″ square iron, filed into an egg-shaped point are screwed to the upper ends of the core. The bobbins are wound full of No. 14 B. & S. gauge cotton-covered magnet wire, if they are to be operated on a battery current of from eight to ten volts such as a plunge battery. If a direct current of 110 volts is available No. 24 should be used. If the rectifier described at the end of this chapter is used No. 22 wire should be used as the voltage of the rectified current on 110 volts alternating is about 90 volts. The magnets must be thoroughly saturated in order to give the best results.

After the magnets are finished they should be fastened to the hard rubber base, at right angles to the rods carrying the spark balls, by two screws through the iron yoke. The balls of the spark-gap should be between the projections of the magnet. A sheet of mica is bent around the polar projections of the magnets in order to prevent the spark from jumping to the cores of the magnets. Fig. 17.

The ends of the coils are so connected that the current will traverse them in opposite directions. The outer terminals are brought out to suitable binding-posts and the interrupter is finished.

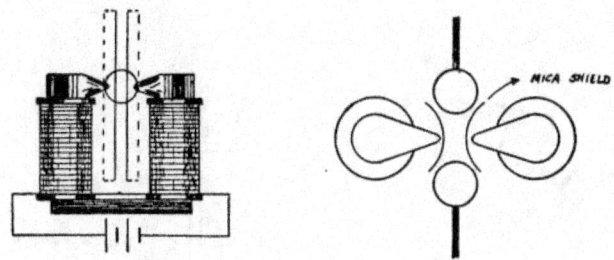

FIG. 17. — MAGNETIC INTERRUPTER.

The principle of this piece of apparatus is based on Davy's experiment in which he found that the electric arc is extinguished upon the approach of a magnet.

THE MOTOR-DRIVEN INTERRUPTER

This interrupter is the one the writers used in their earlier experiments with the 7″ standard coil described in the latter part of this book. It consists essentially of a fan motor run on the alternating current at 110 volts, driving a brass disc having a number of projections bolted around its face, and a brass oscillator so mounted that the distance separating it from the projections on the disc can be varied at will. The motor may be of any suitable design that the builder may possess. A small battery motor running on direct current can be pressed into service if the amateur does

not care to go to the expense of purchasing a fan motor or
has not the facilities for building one. The directions for
building a suitable induction motor are given at the end of
this chapter.

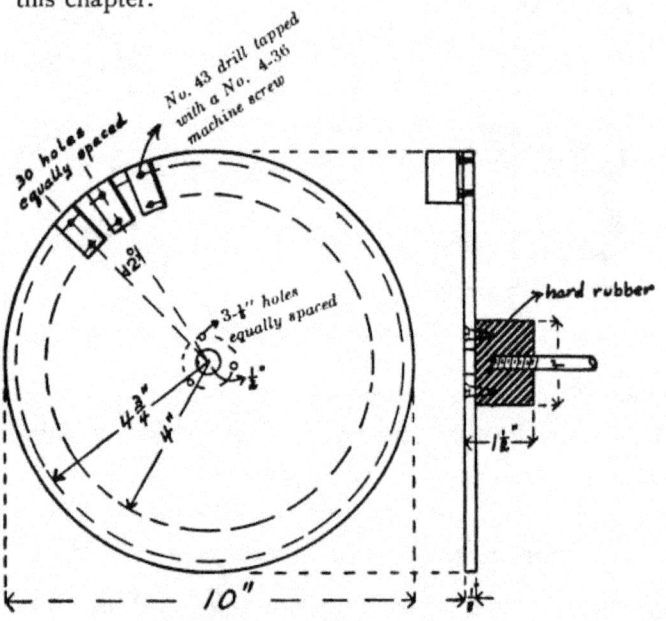

FIG. 18. — MOTOR INTERRUPTER FAN.

To make the disc for this interrupter turn out of $\frac{1}{8}''$ sheet
brass a circular piece 10″ in diameter. If no lathe is available it may be procured at a model maker's quite reasonably.
Lay off on its face two concentric circles, 8″ and $9\frac{1}{2}''$ in diameter respectively. Divide the inner of these into thirty
equal divisions and draw radial lines from the centre of the
disc through each of the points marked off, thus dividing

the outer circle into the same number of equal divisions. Drill a hole through each of the points laid off on both circles and tap them to fit a standard 4–36 machine screw. A number of brass angle pieces made by bending $\frac{1}{16}$" brass

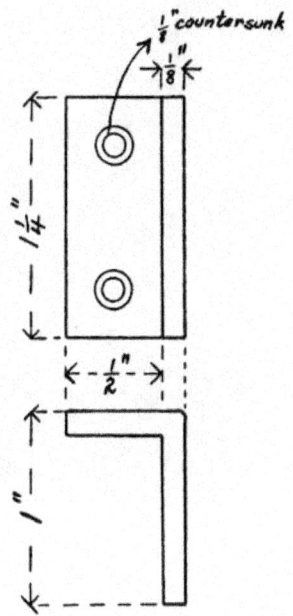

Fig. 19. — Brass Angle Piece.

into the form shown in the figure are procured. Two holes are drilled and tapped in each one to fit a standard 4–36 flat-headed machine screw. These pieces are screwed to the brass disc with $\frac{1}{4}$" screws.

A $\frac{1}{2}$" hole is drilled in the centre of the disc and three $\frac{3}{16}$" holes are drilled on a circle having a radius of $\frac{3}{4}$". Next turn out a circular block of hard rubber 2″ in diameter and the same shape as in Fig. 20. The brass disc is screwed to it with three brass wood screws ¼″ long and the whole is fastened to the shaft of the motor so as to be well insulated from it. To make electrical contact with the brass

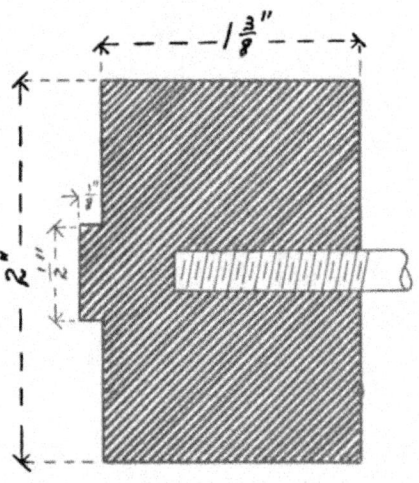

Fig. 20. — Hard Rubber Block.

plate a brush to bear on the back near the centre is cut out of a piece of $\frac{1}{16}$″ sheet spring brass. This piece should be 10″ long and ½″ wide. It is mounted on a piece of hard rubber with a suitable binding-post, so as to press against the back of the disc.

For the other side of the spark-gap a standard mounted on hard rubber similar to the one described for the simple spark-gap is used, but instead of being 4″ long it is 6″ in length and the brass ball is best if it is slightly less than $\frac{1}{2}''$ in diameter. The fibre handle and shield are necessary in order to adjust the length of spark-gap while the coil is in operation.

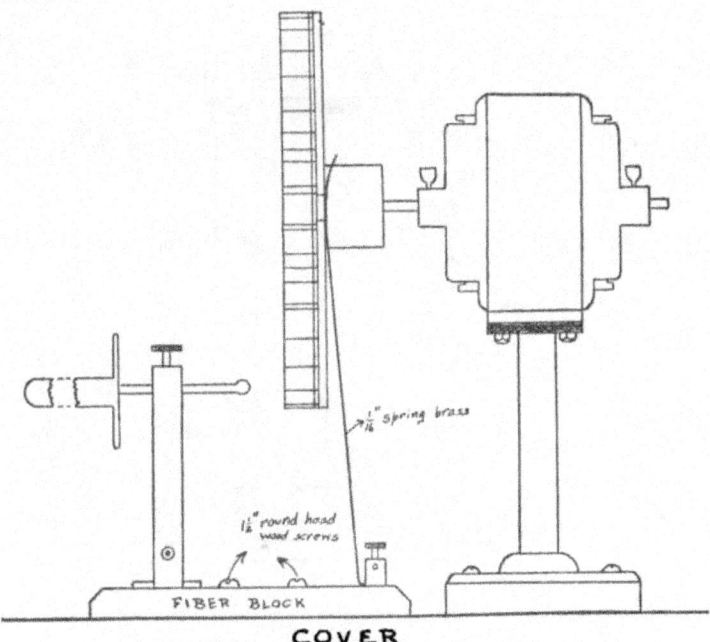

$\frac{1}{16}''$ spring brass

$1\frac{1}{2}''$ round head wood screws

FIBER BLOCK

COVER

FIG. 21. — SECTION OF THE MOTOR INTERRUPTER.

The spark takes place between the brass ball and the projections on the disc. As there is a considerable air current set up by the rapid rotation of the disc, very little arc will form across the spark-gap.

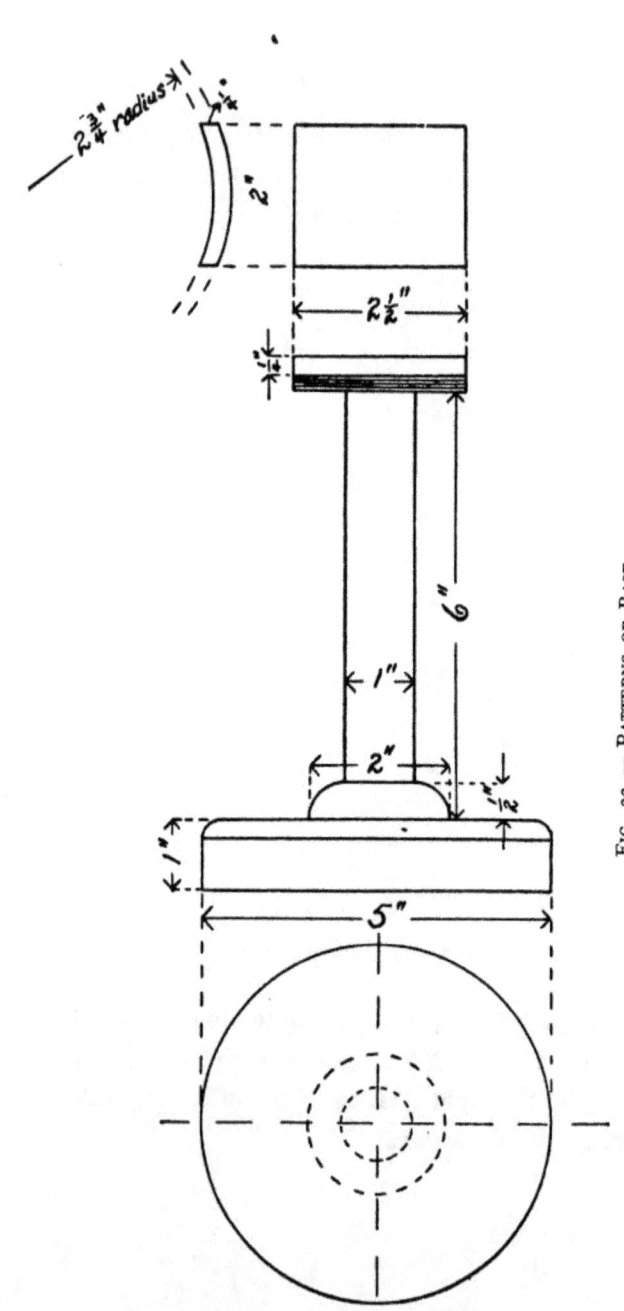

Fig. 22. — Patterns of Base.

A SMALL SELF-STARTING SINGLE-PHASE INDUCTION
MOTOR

To build a motor, to run on the single-phase alternating current of 110 volts, suitable for running a fan interrupter, is perhaps very difficult. The builder will require more tools and a much greater knowledge of machine shop practice to construct an efficient motor than to build all of the parts of Tesla apparatus combined. For those who have had but little experience in motor construction, the writers suggest that the amateur purchase an alternating-current fan motor or a suitable direct-current battery motor. The following description of the building of this motor is given in order that this book may be complete in itself and so that the coil builder will have all the necessary data to build the complete apparatus without reference to other works.

The first step in the construction is to make the necessary patterns for the base and yokes. There are two castings required. The base supporting the punchings for the stator is cast directly on the standard which supports the motor. The drawing, Fig. 23, will give the required dimensions.

It is assumed that the amateur pattern-maker is aware that an iron casting is smaller than the pattern from which the mould was made, therefore shrinkage must be allowed for in the pattern in order to be sure that the casting will be large enough. One eighth of an inch to the foot is about the proper amount.

Turn out a circular piece 5″ in diameter and 1″ in thickness. Then a rod 1″ in diameter, and 6″ long, swelling into a graceful enlargement at the lower end, is turned out and fastened with glue and nails to the centre of the circular disc.

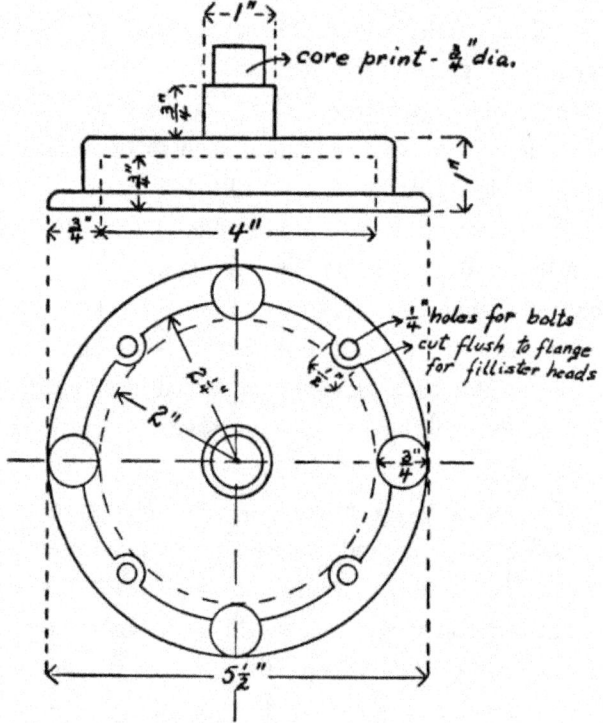

FIG. 23. — PATTERNS OF YOKE.

A piece of wood is cut out of 1″ stock to the form shown in the figure. The radius of curvature of the arc must be $2\frac{3}{4}$″ so that the stator will fit it accurately. This piece is glued and nailed to the top of the upright. The whole pattern is

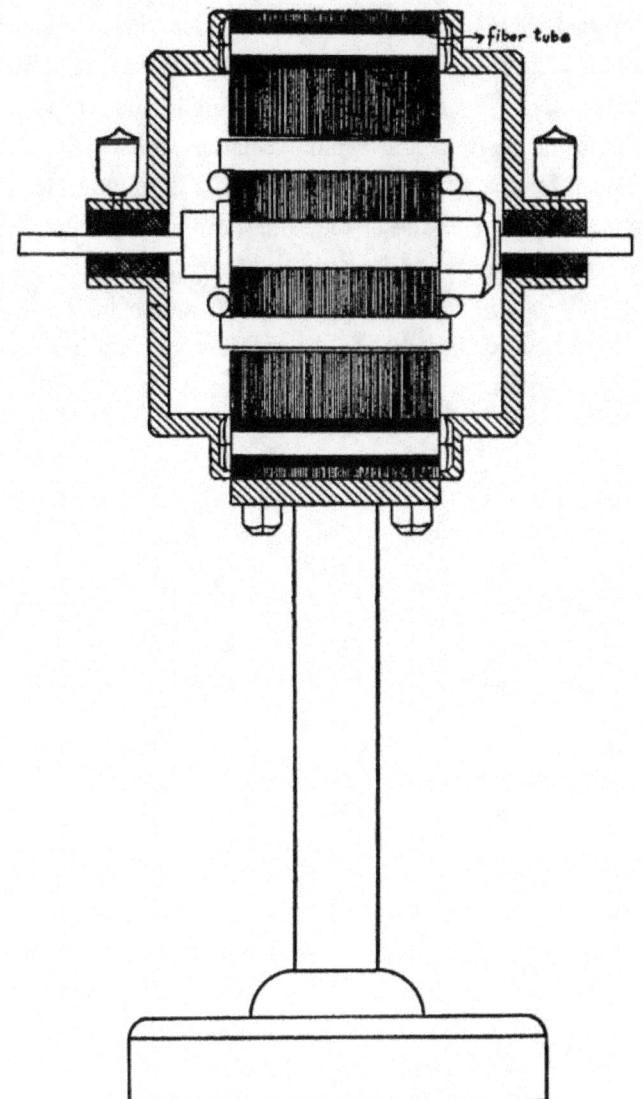

FIG. 24. — SECTION OF COMPLETED MOTOR.

given two coats of best shellac varnish, containing sufficient lampblack to make it jet black. This completes the pattern for the base. In order to provide a support for the bearing of the rotor shaft we must make a pattern for a yoke.

To make this pattern we turn out of $1\frac{1}{2}''$ stock a circular disc $5\frac{1}{2}''$ in diameter and of the same form as in the drawing. Two castings are made from this pattern, one to fit each end of the stator. They not only serve to furnish bearings for the rotor, but also to enclose the entire motor and thus keep out moisture and dust.

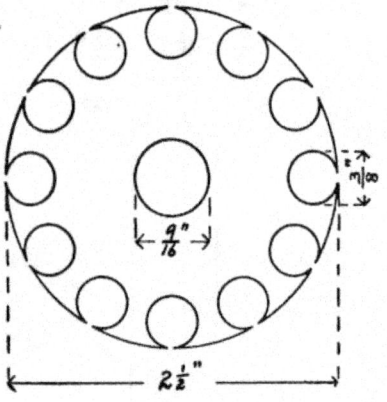

FIG. 25. — ROTOR DISC.

After the patterns are shellacked they should be sent to a foundry where the castings can be obtained quite reasonably. It requires one casting of the base and two of the yoke or journal. When the castings are obtained they should be chipped and all the roughness filed off.

The rotor consists of a number of iron discs $2\frac{1}{2}''$ in diameter and having twelve $\frac{3}{8}''$ holes drilled around the edge and a $\frac{9}{16}''$ hole in the centre. They should be made of iron about $\frac{25}{1000}''$ in thickness; about forty of them making a pile $1''$ high. These discs can be made by the coil builder with the help of a lathe and drill press, or they can be obtained already stamped out from any of the large dealers in electrical supplies. A sufficient number of them are mounted on a shaft, turned down from a $\frac{3}{4}''$ rod of cold rolled steel to the size shown in the figure, to make a pile $2\frac{1}{2}''$ in height. As the motor will not have a very heavy load thrown on it, it will not be necessary to key them to the shaft. A good driving fit is sufficient to keep them from turning. They can be clamped in position by the nut shown in the figure. The conductors consist of twelve $\frac{3}{8}''$ copper rods $2\frac{3}{4}''$ long. One of these rods should be driven in each of the holes around the edge of the discs and should project $\frac{1}{8}''$ beyond them on both sides. To short-circuit them, two heavy rings made by bending two pieces of $\frac{1}{4}''$ copper rod into a circle having an outside diameter of $1\frac{3}{4}''$ are soldered to the ends of the rods. Use sufficient solder in making these connections in order to prevent heating at the connections by the induced current in the rotor. This completes the rotor.

The next thing to consider is the stator punchings. In this case they will not be punched out, but will be cut out on either a shaper or milling machine, or cut out by hand after as much metal has been removed as is possible by drilling. The diameter of these discs is given in the drawing on Fig. 27. To make them, cut roughly out of $\frac{25}{1000}''$

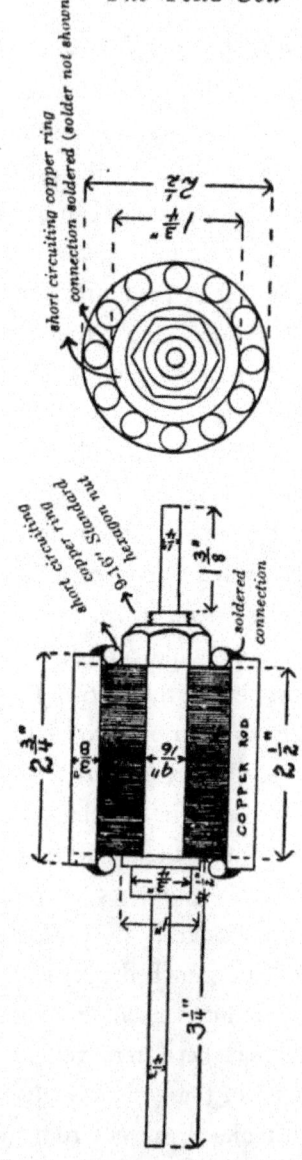

FIG. 26. — ROTOR AND CLAMP NUT.

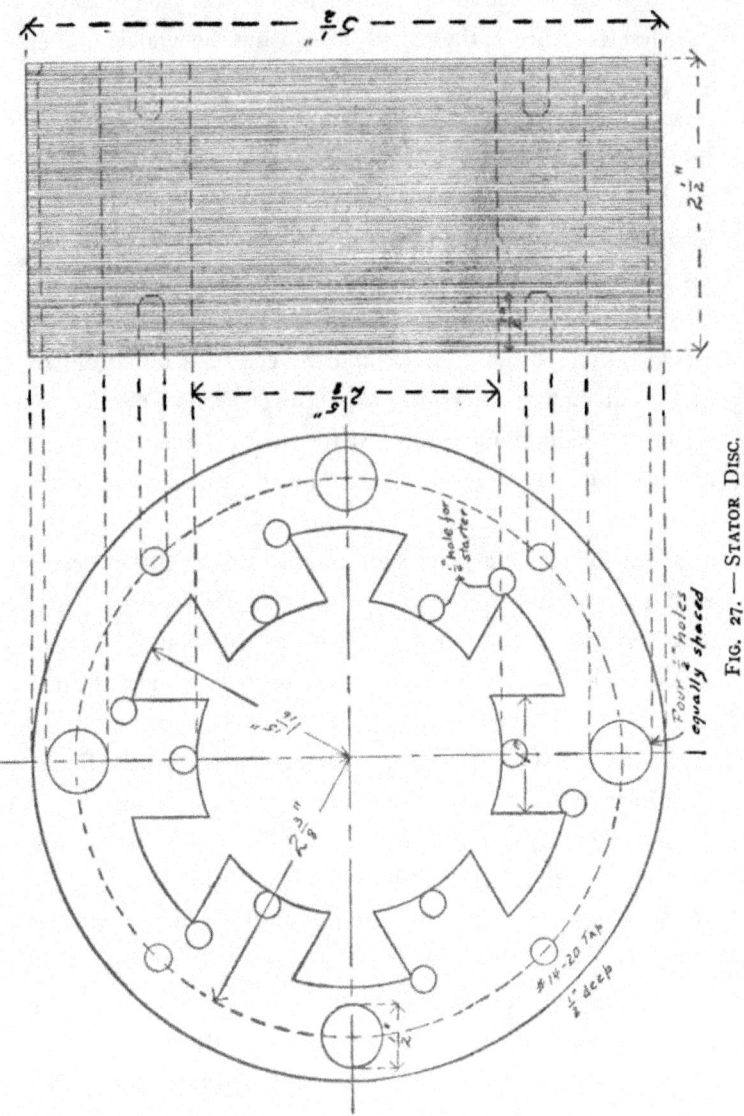

FIG. 27. — STATOR DISC.

iron about 100 pieces 6″ in diameter, with a pair of snips. A ¾″ hole is drilled in the centre of them and the whole number are clamped on a mandrel between two nuts and turned down in a lathe to 5½″. Next four ½″ holes are drilled as in the figure, and a fibre tube ½″ in external diameter and ¼″ internal diameter is driven in each hole. A ¼″ stud 3¼″ in length, with hexagonal nuts and ¾″ iron washers, binds the discs together. After tightening up the nuts the bolts can be slightly riveted to guard against possible loosening. The whole is clamped in a chuck and the centre is bored out to 2⅝″ in diameter. Next the slots are cut as in the figure, either on a milling machine or shaper or by the use of a hack saw and file. A large bulk of the metal can be removed by drilling.

When the stator is complete it is mounted on the pedestal with four cap screws which screw into the bottom edge. The two ends are fastened to the stator with four ½″ fillister screws 1″ long. The holes for these screws are drilled midway between the nuts binding the stator discs together. In order that the heads fit up against the stator four holes should be drilled to allow the nuts to project into the heads.

The rotor is next wrapped with paper until it just fits into the stator and the heads are bolted on in the way they are to be permanently. It is well to mark them so that they can always be put back in the same place. Then the space between the shaft and the journal is filled with the best grade of Babbit metal obtainable. Cardboard washers slipped over the shaft prevent the babbit from running out.

The next thing to do is to wind the stator coils. The wire used is No. 22. The coils are wound in a wooden frame of the size shown in Fig. 28. After the coils are wound they are wrapped with tape, shellacked, and allowed to dry by thoroughly baking. Before the coils can be put into place, means must be provided for making the motor self-starting. This is accomplished by means of a short-circuited copper

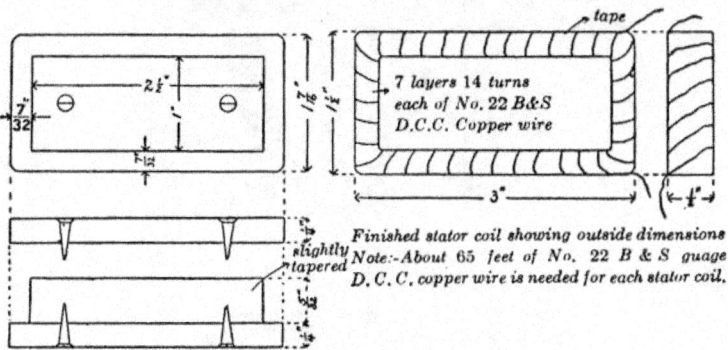

tape

7 layers 14 turns
each of No. 22 B&S
D.C.C. Copper wire

$2\frac{1}{2}$"

3"

Finished stator coil showing outside dimensions
slightly
tapered Note:-About 65 feet of No. 22 B & S guage
D. C. C. copper wire is needed for each stator coil.

Frame on which stator coils are wound.

FIG. 28. — FRAME FOR STATOR COILS.

conductor lying in the grooves marked "A" in the drawing of the stator punchings, Fig. 29. This conductor consists of a piece of No. 14 bare copper wire bent into the form of a rectangle, so as to fit around the one half of the polar projections of the stator. The two ends are soldered together. A glance at the figure will make this clear.

The coils are next slipped into place over these short-circuited conductors. The terminals of the stator coils are so connected as to induce opposite poles in adjacent polar

PLATE IV. — Motor-driven Interrupter.

PLATE V. — The Electrolytic Rectifier.

pieces. The six coils are in series, the end terminals being brought out to suitable binding-posts fastened to the end pieces and suitably insulated from it. A coating of black paint completes the motor.

Although the induction motor is a constant-speed motor at varying loads, we can secure some slight speed regulation,

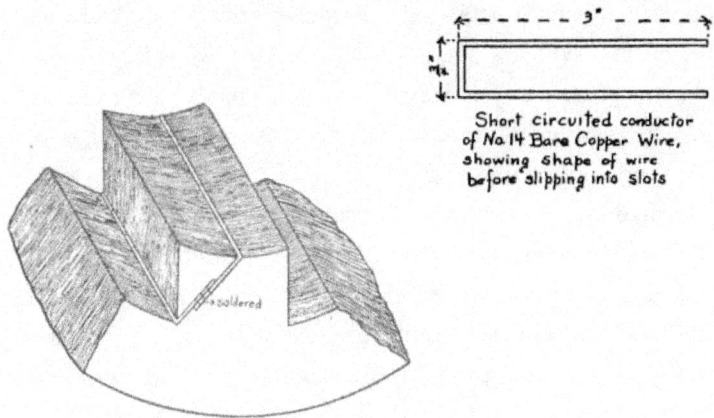

Short circuited conductor
of No. 14 Bare Copper Wire,
showing shape of wire
before slipping into slots

FIG. 29 — SELF-STARTING DEVICE.

which is a great advantage in operating the coil, by interposing a liquid resistance in series with the motor. This resistance consists essentially of any suitable glass jar, such as is used in a Daniell or Gravity cell, having two metal plates suspended in an electrolyte, so that the distance between them can be varied at will. Copper sulphate is generally used as the electrolyte.

AN ELECTROLYTIC CURRENT RECTIFIER

It was only at the last moment that the authors decided to make public the results of their experiments on an electrolytic current rectifier, which has proven highly satisfactory. Its advantages are that it is easily and cheaply built, it requires only slight attention, its efficiency is very high, and the current which a small set will rectify is very large.

To a great many it may seem out of place in putting in this description of a rectifier, which is entirely foreign to the Tesla apparatus. The reasons for so doing, however appeared, to the authors at least, sufficiently great, for if the amateur constructs the magnetic wipe-out he will need a source of direct current at about 80 or 90 volts pressure, since this current can hardly be obtained from the lighting circuits which are generally alternating or from batteries. Then having this source of direct current he will be able to substitute a D. C. motor for the induction motor described in this chapter.

The greater number of rectifiers now on the market use the method of choking out the one half of the alternating-current wave and it is to this fact that their low efficiency is due. The high efficiency of the apparatus devised by the authors depends on what might be called the alternate path connection or method; that is, there are two paths for the current to traverse, one of enormous resistance and one of very low resistance.

The idea of this form of rectifier came to the authors in

the following way. They were experimenting on some cathode tubes of peculiar construction, using a 12″ induction coil. The current from the secondary of this coil is oscillatory in character, of course. It was observed that the discharge through the tube was very unsteady, especially when the interruptions were not very rapid. A line of experiments was carried out to determine the cause of this unusual effect, with the result that the resistance was found to be enormously greater for currents in the one direction through the tube than in the opposite direction, due entirely to the difference in the forms of the two electrodes.

After discovering this fact they wondered if some electrolytic cell might not be made which would possess the same properties and could be used to rectify the ordinary alternating currents. From a previous study of the effects of various electrodes on the electrolysis of certain solutions we arrived at several cells which exhibited these properties to a marked extent.

It was found that an aluminium electrode was the essential thing in every cell, together with some acid salt capable of forming an oxide with aluminium. The other electrode might be any conductor unaffected by the solution.

Some of the conductors suitable for the other electrode are iron, carbon, and lead, and the following solutions all gave more or less satisfactory results. Acid sodium carbonate, acid sodium phosphate, acid potassium tartrate, potassium alum, and in fact most of the ionizable, slightly acid sulphates, carbonates, tartrates, and phosphates.

By merely putting one of these cells in the circuit, the one half of the alternating current wave may be choked out. But this method gives an efficiency of less than 50%. Thus the authors were led to devise the alternate-path method.

Before describing this method, however, we will take up in detail the properties of a single cell. After a current is passed for a few minutes through one of the cells, a coating of oxide is formed on the aluminium electrode which is practically a non-conductor. While this does not prevent the difference of potential from being maintained across the cell, it does prevent the ions from giving up their charge and in this way it acts like a polarized copper plate in a single galvanic cell. This condition of enormous resistance exists when the aluminium is the anode. When on the reversal of the current the aluminium becomes the cathode there is merely the resistance of the electrolyte encountered. Any cell possessing this property is called asymmetric.

As stated before, a single cell by being merely put in the circuit would choke out the one half of the alternating wave, but as this gives an intermittent current, the following is the method devised by the authors. Three cells are needed in all. Two of these consist of one electrode of aluminium and one of iron, with a solution of sodium acid carbonate. The third has two aluminium plates and one iron plate between them. The same solution is used.

On looking at Fig. 30 it will be seen that when E is positive the current can flow from either plate 2 or 3 across the electrolyte to plates 1 or 4. The path from 3 to 4 is of enormous resistance, as the aluminium is the anode, but the path from 1 to 2 is of low resistance and hence the current takes this path. When *H* becomes positive the current can flow from 6 to 7 or from 5 to 4. It takes the path from 6 to 7 as this is of low resistance. In this way both waves of the alternating current are used and the only loss is due to the resistance of the electrolyte.

Thus the direct current from this set has a sine wave form, in which all the negative values in the alternating have been made positive.

The following are plans for a rectifier suitable for use directly on the 110-volt-alternating-current light mains. The rectifying cells have glass containing jars. The jars are all 7″ x 6″ x 4″ inside measurements. The aluminium plates are cut out of ⅛″ sheet and are all the same size and shape. They are 5″ x 7″ and four are required. The aluminium should be comparatively pure to prevent deterioration of the plates due to local action. If the plates are to any extent impure the cells may fail to work, and if they do rectify it will be at a very low efficiency. The iron plates are cut from ⅛″ sheet. Two of them are the same size as the aluminium plate and the third is 8″ x 5″. This larger plate is to be used in the middle cell. The necessity for making it larger is that it goes between two aluminium plates and the extra length is required to fasten the binding-post to. The plates are held three eighths of an inch apart in the following manner: Out of some ⅜″ sheet vulcanite or hard rubber (fibre must not be used as it swells in water) cut four strips $\frac{1}{2}'' \times 6\frac{1}{2}''$. Also cut out four washers about $\frac{1}{2}''$ in diameter and a number of pieces $\frac{1}{2}''$ square. These latter pieces are drilled and tapped to fit a standard $\frac{1}{4}''$ thread. The washers have a $\frac{1}{4}''$ hole drilled in them. Some $\frac{1}{4}''$ vulcanite rod is cut up into about $2''$ lengths and threaded to fit a $\frac{1}{4}''$ nut. With these strips and washers the plates are held the required distance apart and the bolts firmly fasten them together. The strips of rubber are used across the tops of the plates and the washers at the bottom. See Fig. 30. This method of using hard rubber bolts and nuts is far superior to using iron ones and fitting them with an insulating bushing. A binding-post is fastened at the top of each one of the plates. As the strips across the top of the plates are longer than the jars are wide; when the electrodes are put in place they will be suspended in the solution by the strips resting on the edges of the jars.

The electrolyte, if sodium acid carbonate is used, should be a saturated solution. Other solutions than this can be used, although the authors obtained the best results with this one. Besides it is about the cheapest of all the possible electrolytes.

In selecting an electrolyte the following factors must be taken into consideration. It must have low resistance, it must be a stable compound, and when no current is flowing it must not attack the aluminium plate and only slightly attack it when current is passing.

To make the set convenient to handle the jars should be mounted in a wooden frame, with the cell containing the

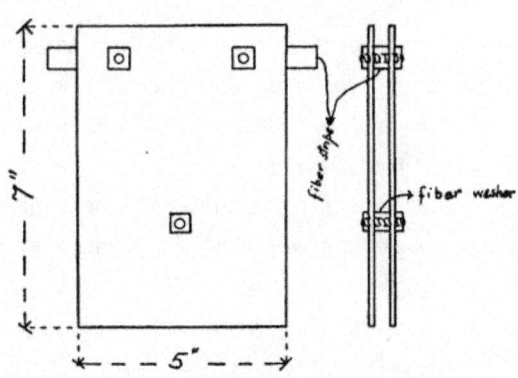

Method of forming electrodes

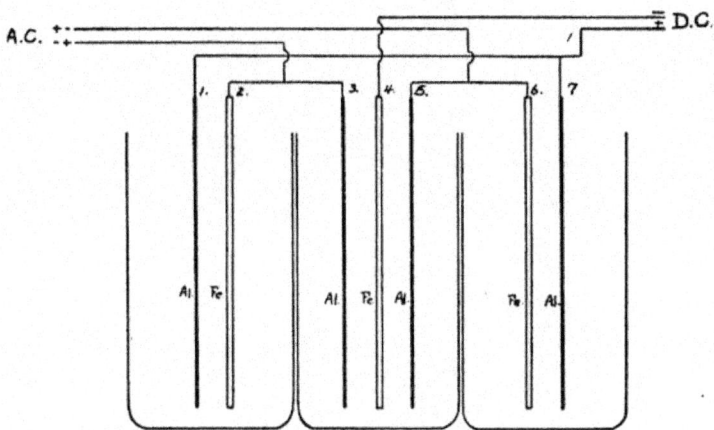

Figure showing position of the cells and electrodes with the connections. Fe.-Iron   Al - Aluminium.

Fig 30. — Rectifier Plates and Wiring Diagram.

two aluminium electrodes in the middle. The connections are shown in the figure.

With this electrolyte the aluminium plates will form in a few minutes, it being merely necessary to short circuit the D. C. taps with a resistance and allowing the rectifier to take a full load current.

The efficiency of the apparatus will be somewhat increased by using a cooling worm, as the electrolyte when cool has the greatest current density.

It will be necessary to renew the electrolyte at intervals, depending on the use the set is given

CHAPTER VI

THE CONSTRUCTION OF THE BOXES

THE next thing to consider in the building of a Tesla coil is the boxes which contain the transformer and high-tension coil. One box for transformer, condenser, and high-tension coil might be used, but for a coil of this size the weight would be objectionable. Two separate boxes give the ideal result. They have the advantage of not being too bulky to handle, and the transformer in this form can be used separately, if so desired. A single box, however, has the advantage of taking up less room and of having all the high-potential connections inside where they are safe, except those which lead to the interrupter which is placed on top of the box.

Oak makes the most substantial box, but it is harder to make tight, owing to the fact that the shellac varnish which is used for filling up the pores in the wood does not sink into oak with the same readiness as it does in a softer wood. Pine is the best material to use as the joints will require considerable filling up to make them impervious to paraffine oil, which will soak through almost anything in time. Sugar pine may be readily stained and looks very neat when varnished.

For the sides and ends of the transformer box it requires

a piece of straight-grained pine free from knots, $1\frac{1}{2}''$ x $10''$, $6''$ long. The bottom should be made of a piece of $1\frac{1}{2}''$ x $14''$, $26''$ long, and the top of a piece of $1''$ x $11''$, $2'$ long. Cut the pieces to the size shown in Fig. 31, and plane the edges true. The end pieces must be mortised into the sides $1''$ from the end. These tongues and grooves may be cut with a saw and chisel if a rabbeting plane is not at hand. After the sides

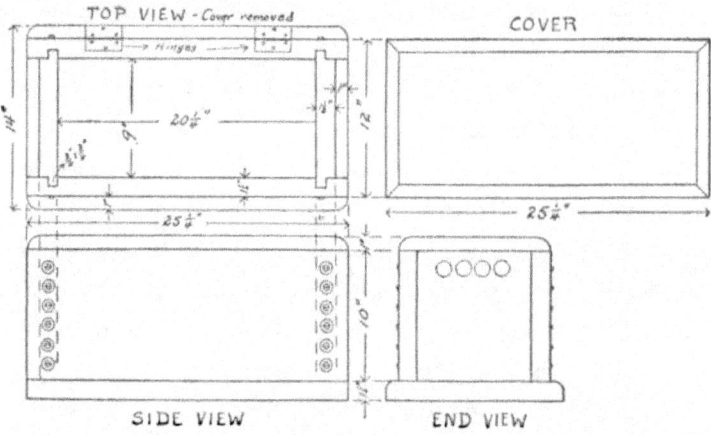

TOP VIEW - Cover removed   COVER

SIDE VIEW   END VIEW

FIG. 31. — TRANSFORMER BOX.

and ends are finished the tongues and grooves are given a heavy coat of shellac, which has been dissolved in grain alcohol, and while still wet are put together. Six long brass screws with round heads are to be used in each board to hold the sides. A brass washer $\frac{1}{2}''$ in diameter should be placed on the screw to prevent the head from sinking into the wood.

Next the edges should be gone over with a plane if necessary so that the bottom board will fit flush in all places. The bottom board is to be 1" wider than the width of the box, so that it laps over ½" on each side. When the bottom board is cut to size, the edges are rounded off with a plane to give a finish, and then it is fastened to the box with long flat-headed brass screws, placed every four inches along the sides and ends. A coating of shellac is given to the edges of the box just before putting the bottom on, to help make it tight. The screws must be forced in until they are flush with the wood.

Next the inside should be given five or six coats of shellac, paying especial care to get it into the joints, and allowing each coat to dry before applying the next.

A small brass cock in the end near the bottom is a convenience in emptying the box of its oil, but the labor of putting it in so that the box will not leak is such that a siphon is quick enough for an occasional emptying of the oil.

The box for the high-frequency coil and condenser must have the same care taken in its construction as in the case of the transformer box. The dimensions are given in the working drawings in Fig. 32. A partition is put in between the condenser and oscillation transformer, but several holes should be bored in it near the bottom to allow of the free circulation of the oil. This box must also have several coats of shellac, as the insulating oil used will leak through in spite of all the precaution taken.

After the boxes are finished they should be stained or varnished to suit the taste of the builder. Walnut stain looks well, and as it is dark it covers up a multitude of faults in the wood working. If the boxes are well made a good oiling followed by several coats of shellac makes a very good finish.

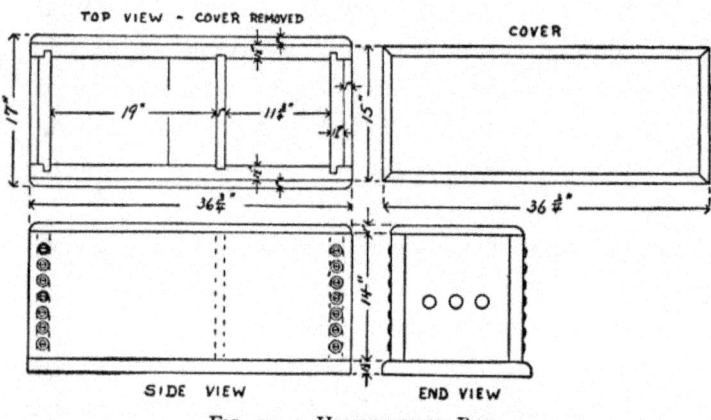

FIG. 32. — HIGH-TENSION BOX.

Everything is now ready for the assembling of the parts, which will be taken up in the next chapter.

CHAPTER VII

ASSEMBLING

IT is not wise to hurry when assembling the apparatus, for if the high-tension wires are not properly insulated, brush-discharge effects will be noticed on operating. In nine cases out of ten poor insulation will result in puncturing his condenser and probably burning out his transformer. Care should be taken to follow these directions.

First mount the transformer in its box. After lowering the transformer into its box bring its four primary leads to four heavy binding-posts on the end of the box. The two inner terminals of the two sections are brought to two adjacent binding-posts, and the two outer ones to the other two, in such a manner that connecting electrically the two middle binding-posts puts the sections in series and short-circuiting the two outer pairs throws the sections in parallel. See the diagram. This is accomplished by means of a piece of flat brass with slots filed in it so that it just fits across two binding-posts, or by a short piece of brass rod which fits in the holes of the binding-posts.

The secondary terminals are soldered directly to two brass rods $\frac{3}{8}''$ in diameter and $3''$ long, which extend through the opposite end of the box for $1\frac{1}{2}''$. These rods are insulated

by two heavy hard rubber or fibre bushings made as follows: A 2″ piece of hard rubber or fibre rod 2″ in diameter
is turned down to 1½″, except for a ⅜″ flange on one end the
full diameter of the rod. A ⅜″ hole is drilled down the centre
of the bushing. The bushings should be a driving fit both

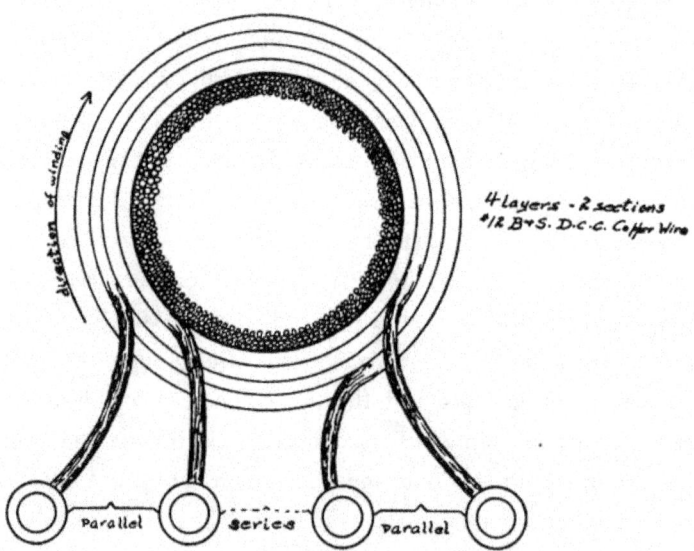

FIG. 33. — CONNECTIONS FOR PRIMARY OF TRANSFORMER.

in the end of the box and over the brass rod. The rods
are 6″ apart and 8″ up from the bottom of the box. This
brings them well above the level of the oil, thus assuring no
leakage at this point.

The leads from the sections should be led to the brass
rods through glass tubing bent in the desired shape. A hole
is drilled and tapped in the end of each rod to fit a standard

brass set screw. Another hole is drilled at right angles to the first about $\frac{1}{2}''$ from the end of the rod to meet the first hole. This makes an efficient binding-post to hold the conductor.

This finishes the connections on the transformer, which can now be placed in the position which it is to occupy. The box is filled with enough pure paraffine oil to cover the transformer. This oil should be of the best quality obtainable, free from moisture and impurities, such as is used for insulating purposes. It should be allowed to soak into all the sections for 24 hours before using. The primary terminals are connected to a source of alternating current by means of a suitable switch and fuse capable of carrying 30 amperes.

The next step is putting in the connections and terminals in the high-tension box.

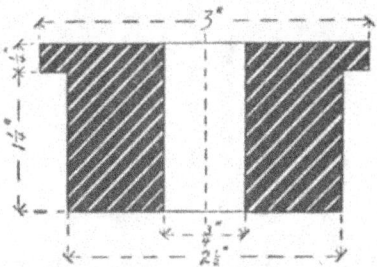

FIG. 34 — HIGH-TENSION BUSHING.

Three brass rods similar to those used in the transformer box, with the same form of bushings, are driven through holes in the end of the high-tension box next to one end of the oscillation transformer. These holes should be about 6″ from the bottom and 3″ apart. The brass rods project

¼″ within the box. The bushings and rods can be made oil tight by giving them a good coating of Le Page's glue before driving them into place.

A ½″ strip of wood is glued on in the lower inside end of the box below these rods, to prevent the end of the oscillation transformer from coming in contact with them.

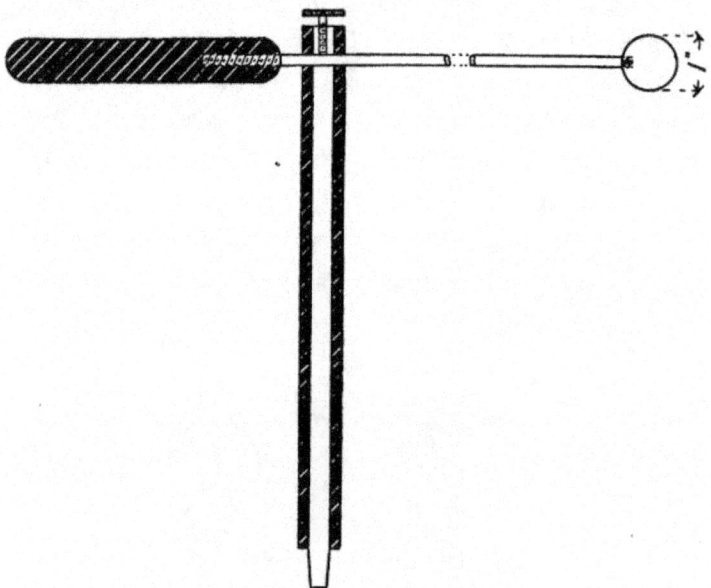

FIG. 35. — OSCILLATORS AND STANDARDS.

The condenser is first lowered into place in the box and a wire is run from one terminal to the nearest outer brass rod, to which it is soldered. The wire should be enclosed in a glass tube suitably bent and should follow the lower edge of the box. A wire is run from the other terminal of the condenser in a similar manner to the other outer brass rod. It follows the other lower edge of the box. A tap wire is led from this conductor at a point opposite to where one terminal of the primary band of the oscillation transformer is to be. A wire is also soldered to the middle brass rod. The oscillation transformer is now lowered into place. The tap wire from the condenser lead is soldered to one end of the primary band and the wire from the middle brass rod to the other end of the band. These leads should be run in glass tubes and as directly as possible. They should also be kept under the oil. The connections are shown in Fig. 36.

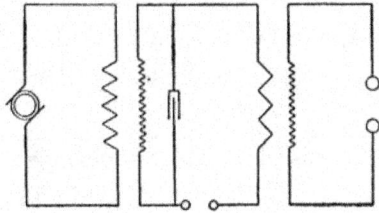

FIG. 36. — WIRING DIAGRAM.

After all connections have been securely soldered the box is filled with oil, so that the entire apparatus is completely immersed.

The oscillators and standards can now be constructed. Two fibre or hard rubber bushings $2\frac{1}{2}''$ in diameter and $1\frac{1}{2}''$ in length, and having a flange $\frac{1}{4}''$ thick and $3''$ in diameter turned on one end, are set in two holes cut in the cover directly above the holes in the brass bushings on the oscillation transformer. A $\frac{3}{4}''$ hole is drilled through the centre of each bushing. Two $\frac{3}{8}''$ brass rods 10" long are enclosed in fibre tubes $\frac{3}{4}''$ in outside diameter and $9\frac{1}{2}''$ long. The tubes should fit the rods tightly. The ends of the brass rods project from the fibre and can be slightly tapered to fit the bushings on the oscillation transformer.

The oscillators consist of two brass balls 1" in diameter screwed on the end of two $\frac{3}{16}''$ brass rods 12" long, which are to slide easily in two holes drilled $\frac{1}{2}''$ from the top of the standards through both the fibre and the rod. A set screw at the top of each standard will be of convenience in clamping the rods in any desired position.

In order that the discharge gap may be adjusted while the coil is in operation two vulcanite handles $\frac{3}{4}''$ in diameter are screwed on the ends of the rods, carrying the oscillators, for about $1\frac{1}{2}''$.

Slide the standards through the bushings in the cover until the rods make good contact with the bushings on the oscillation transformer. This completes the connections in the second box. It should now be placed in its final position at the high-tension end of the transformer, leaving a space of about $1\frac{1}{2}'$ for the interrupter between the two boxes.

The particular form of interrupter which has already been built is connected to the binding-posts of the high-tension coil as in the wiring diagram. These leads and those from the transformer to the high-tension box should be of No. 12 B. & S. gauge hard copper wire and enclosed in the heaviest glass tubing obtainable. They should be as straight and as short as is consistent with safety in operating the primary spark-gap.

A suitable panel switchboard, with the necessary fuses, transformer and interrupter switches, makes a desirable acquisition to the apparatus, but it is not essential. This matter is left to the taste of the individual worker.

When the connections have all been made and the oil has driven out all the air that it can, open the interrupter gap about 1″ and cautiously close the transformer switch. If no excessive load is taken by the transformer as manifested by the 30 ampere fuses or an alternating-current ammeter in series, if one is obtainable, the spark-gap can be slowly closed until the condenser discharges across it. Then, if the directions have been carefully followed in building the apparatus, a heavy, bluish white, snapping discharge of over 12″ in length will pass between the oscillators, upon the further adjustment of the interrupter gap.

When the transformer is used on 55 volts the primary sections should be connected in parallel and for 110 volts in series. If the sections on the primary were connected in parallel on 110 volts, the voltage output of the secondary would be about double what it ought to, and hence the condenser may puncture.

If by any chance a discharge should fail to take place, the fault may be due to several things. In most cases it will be due to the fact that the sections on the primary or secondary of the transformer are connected in opposition. To determine whether the transformer is working satisfactorily or not, disconnect it from the rest of the apparatus and see if an arc discharge of at least 6 or 8 inches can be drawn out between two electrodes. This arc is generally of a yellow color and easily extinguished by any draught of air. If you do not obtain a 6 or 8 inch arc test the sections to see if they are connected in the right manner, and if they are and no arc still results, which is highly improbable, then some error has been made in the construction.

If the fault does not lie in the transformer, it is most likely that it lies in the condenser. To test this connect the transformer up with the condenser and see if a condenser discharge, determined by its bluish white color, can be obtained in the primary spark-gap. If there is none obtained, your condenser is most likely short-circuited or even punctured, which can only be remedied by its reconstruction.

The next place to look for trouble is in the oscillation transformer. Ring out by means of a magneto the primary and secondary circuits to see that there are no open circuits, and then see if there is any short circuit between the primary and secondary. If these parts are all right the fault may be due to poor insulation, in having the turns of the secondary touching. The remedy is obvious.

Finally look and see that all the electrical connections are as they should be, then the apparatus cannot fail to discharge over a 15″ air-gap.

CHAPTER VIII

THE THEORY OF THE TESLA COIL

ALTHOUGH, in the introduction, the authors stated that they would not attempt to give a mathematical explanation of the coil, still they feel that a few facts regarding the theory would not be out of place here, in that it may suggest certain improvements to the reader. It will also be of assistance if the amateur wishes to construct a coil of his own design.

The first thing to consider is the transformer. Its action, as is well known to almost everybody, depends on electro-magnetic induction. The alternating current flowing in the primary sets up an alternating magnetic field, which being linked with the secondary induces an electromotive force in it. When the secondary is open there is theoretically no current passing through the primary, due to its high self-induction, except that necessary to magnetize the core. As a load is thrown on to the secondary, the current through the primary automatically adjusts itself as the self-induction is decreased by the opposing ampere turns of the second-ary, that is, if the transformer is self-regulating for vary-ing loads.

The normal current through the primary of the transformer used in the 12″ coil is from 22 to 25 amperes, the secondary voltage being about seventy-five hundred, and thus the amperage in the secondary is about $\frac{1}{300}$.

To get the required voltage of seventy-five hundred in the secondary on fifty-five volts in the primary it is necessary to connect the two sections of the primary in parallel, as this has the effect of cutting the primary turns in two. On one hundred and ten volts the sections are connected in series.

The use of the transformer in the Tesla apparatus is merely to charge a condenser, and thus it is seen that an ordinary induction coil or even a static machine of the proper dimensions could be used, but they are not nearly as handy.

Another important matter in connection with the construction of a transformer to be used for creating electric oscillations is to secure a sufficiently small resistance in the secondary. The reason for this is that the transformer is used to charge a condenser.

When an electromotive force is applied to the terminals of a condenser, the full difference of potential is not created between the terminals of the condenser immediately, but rises gradually. The time required to charge the condenser depends on its capacity (C) and the resistance (R) of the charging circuit. The product CR is called the time constant of the condenser, and practically the condenser is charged in a time equal to ten times the time constant. The time constant is to be reckoned as the product of the capacity (C) in microfarads and the resistance (R) of the charging circuit in megohms. The time is given in fractions of a second.

The condenser, more than anything else, constitutes the

essential part of the Tesla coil. It plays the same part as the mechanical interrupter in the ordinary induction coil. Its action, however, is purely electrical and its great advantage lies in setting up the currents of enormous frequency.

When any condenser is discharged, the discharge may take one of several forms, depending only on the three electrical constants of the discharging circuit — inductance, capacity, and resistance. The discharge may be either oscillatory or entirely unidirectional, consisting only of a gradual equalization of the potentials on the two plates.

This may be made clear by the following mechanical illustration. Suppose a glass U-tube to be partly filled with mercury, and the mercury to be displaced so that the level in one side of the tube is higher than in the other. There is then a force due to the difference of level, tending to cause the liquid to return to an equal height in both limbs. If the mercury is now allowed to return, but is constrained, so that it is released slowly, it goes back to its original position without oscillations. If, however, the constraint is suddenly removed, then owing to the inertia of the mercury it overshoots the position of equilibrium and oscillations are created. If the tube is rough in the interior, or the liquid viscous, these oscillations will quickly subside, being damped out by friction.

What we call inertia in material substances corresponds with the inductance of an electric circuit and the frictional resistance experienced by a liquid moving in the tube, with the electrical resistance of a circuit. If we suppose the

U-tube to include air above the mercury and to be closed up at its ends, the compressibility of the enclosed air would correspond to the electrical capacity in a circuit.

The necessary conditions for the creation of mechanical oscillations in a material system or substance are that there must be a self-recovering displaceability of some kind, and the matter displaced must possess inertia; in other words, the thing moved must tend to go back to its original position when the restraining force is removed, and must overshoot the position of equilibrium in so doing. Frictional resistance causes decay in the amplitude of the oscillations by dissipating their energy as heat.

In the same way the conditions for establishing electrical oscillations in a circuit is that it must connect two bodies having electrical capacity with respect to each other, such as the plates of a condenser, and the circuit itself must possess inductance and low resistance. Under these conditions, the sudden release of the electric strain results in the production of an oscillatory electric current in the circuit, provided the resistance of the circuit is less than a certain critical value. We have these conditions present when the two coatings of a Leyden jar are connected by a. heavy copper wire.

Professor William Thomson, titled Lord Kelvin, published in 1853 a paper on "Transient Electric Currents" in which the discharge of the Leyden jar was mathematically treated in a manner that elucidated important facts.

If we consider the case of a Leyden jar or condenser charged through a circuit having inductance and resistance, then in the act of discharge the electrostatic energy stored up in the condenser is converted into electric current energy and dissipated as heat in the connecting circuit. At any moment the rate of decrease of the energy in the jar is equal to the rate of dissipation of the energy in the discharging circuit plus the rate of change of the kinetic or magnetic energy associated with the circuit.

From these facts Lord Kelvin sets up an equation of energy, which leads to a certain class of differential equation having two solutions. The solutions in this case depend on the relation between the constants inductance, resistance, and capacity.

If L = inductance, C = capacity, R = resistance, then the solutions are determined by the relative values of $\dfrac{L}{R}$ and LC.

If $\dfrac{R^2}{4L^2}$ is greater than $\dfrac{1}{LC}$, that is, if R is greater than $\sqrt{\dfrac{4L}{C}}$, or if $\dfrac{RC}{4}$ is greater than $\dfrac{L}{R}$, the charge in the jar dies away gradually as the time increases, in such a manner that the discharge current is always in one direction.

The ratio $\dfrac{L}{R}$ is called the time constant (T) of the discharge circuit, and the product CR is called the time constant (T') of the condenser circuit. Hence the discharge is unidirectional when the time constant of the inductive circuit is less than half the geometric mean of the time constants of the inductive circuit and condenser circuit:

If, however, $\dfrac{CR}{4}$ is less than $\dfrac{L}{R}$ the discharge current will be oscillatory, the current decaying in accordance with the law of a damped oscillation train.

When the discharge is so highly oscillatory that the current is not uniformly distributed through the cross-section of the conductor, then the ordinary resistance (R) and inductance (L) must be replaced by the high-frequency resistance and inductance of the circuit.

When the discharge takes the oscillatory form the frequency is given by the expression,

$$n = \frac{1}{2\pi} \sqrt{\frac{1}{LC} - \frac{R^2}{4L^2}}$$

If R is very small, then $\dfrac{R^2}{4L^2}$ can be neglected in comparison with $\dfrac{1}{LC}$, and then the frequency is given by the expression,

$$n = \frac{1}{2\pi} \sqrt{\frac{1}{LC}}$$

In this equation both the quantities C and L must be measured in electromagnetic units or both in practical units, viz., in henrys and farads.

In the majority of cases in which electric oscillations are practically used, the resistance of the oscillatory circuit is negligible, and the inductance is small and hence easily measured in centimeters or absolute C. G. S. units, one milli-henry being equal to a million centimeters (10^6).

Also the capacity is best measured in microfarads; one microfarad being the one millionth part of a farad or 10^{-15} of an absolute C. G. S., unit (electromagnetic) of capacity.

Hence when L is expressed in centimeters and C in microfarads, the expression for the natural frequency of the circuit becomes

$$n = \frac{5.033 \times 10^6}{\sqrt{CL}}$$

The energy storing capacity of a condenser is given by the expression $\frac{1}{2} CV^2$, where C is the capacity of the condenser and V the charging voltage.

The main thing in constructing condensers to be used on high charging voltages is the solid dielectric. There are in all only a few dielectrics suitable for high-tension work, and this number is reduced when cost and internal energy loss in the dielectric are considered. Glass of certain compositions, ebonite, mica, and micanite are practically all that are suitable, and of these flint glass is the best, as its dielectric constant is high, being from 5 to 10, and its dielectric strength is also great. Glass is brittle, however, and liable to have flaws which sooner or later give way.

The capacity of a condenser depends on the area of the plates, their distance apart and the constant of the dielectric used, and is expressed by the following formula in microfarads, where K is the dielectric constant, S the total area of the plates expressed in square centimeters, and D the distance apart in centimeters,

$$C = \frac{KS}{4\pi D \times 9 \times 10^5}$$

The constant 9×10^5 comes from the fact that one microfarad equals 900,000 electrostatic units of capacity.

The oscillation transformer is nothing but a modified transformer with an air core. The only important facts about its construction are that it should be built to withstand great voltage differences between the turns, and that the primary should have as small an inductance as is practicable, in order to make the frequency as great as possible. No advantage is gained by having many close turns in the primary, because the increase of inductive effect on the secondary, due to an increase in the number of primary turns, is about exactly annulled by the decreased current through the primary due to its own greater inductance.

The function of the interrupter is to destroy any arc that may be formed across the terminals of the primary spark-gap, for if this arc is not wiped out there will be no true oscillatory discharge in the condenser circuit or only a feeble one. The reason for this is that as long as the arc discharge continues, the secondary terminals of the transformer are reduced to nearly the same potential, or at most differ only by a few hundred volts.

The function of the primary spark-gap is to regulate the voltage to which to charge the condenser. Since the potential difference between the spark balls is almost equal to the potential difference across the condenser, the condenser will discharge at a voltage determined by the length of the air-gap. Now there is a certain length of spark-gap which is best suited for each coil and it can easily be determined by trial. As a rule it is best to start with a rather short spark-gap, gradually lengthening it out until a point is almost reached, when opening it out any further would cause it to cease passing. This spark length almost always gives the best results.

In the earlier part of this chapter it was stated that the high-frequency resistance and inductance should be substituted for the ordinary resistance and inductance, when dealing with circuits which are subject to the action of electric oscillations. The processes and means used for the measurement of low-frequency alternating currents and potentials are not always applicable or correct either when applied to high-frequency measurements. The main reason for the difference between the two cases is to be found in the fact that a high-frequency current does not penetrate into the interior of a thick solid conductor of good conductivity, but is merely a surface or skin effect.

When traversed by an alternating current, there are five qualities of a circuit to be considered.

1. The resistance of the conductor, which is always greater for high-frequency currents than for the ordinary currents; that is, direct currents and alternating currents up to about a frequency of 100 per second.

2. The inductance of the conductor depends on its geometrical form, material, and the nature of the surrounding insulator. The greater the frequency, the smaller the inductance becomes.

3. The capacity of the conductor, depending on its position with regard to the return circuit and other circuits and on the dielectric constant of the surrounding insulator.

4. The dielectric conductance of the insulator surrounding the conductor.

5. The energy dissipating power, due to other causes than conductance, such as dielectric hysteresis, which exist in the dielectric. Under this heading comes the loss of energy from the brush discharges through the air between the conductors.

If the constants of a circuit for low-frequency currents are known, the values of the constants for high frequencies can be calculated fairly correct. The high-frequency constants can, however, be measured directly, but the apparatus is rather delicate and inconvenient for use and besides not always satisfactory. If the coil builder cares to measure the constants of a circuit for himself, he will find the description of the necessary instruments in other books as it is beyond the scope of this work.

Having now briefly treated theoretically on all four of the principal parts, the authors will try to show how these parts work together to form the Tesla high-frequency apparatus.

The condenser is connected in series with the secondary of the transformer and thus is being continually charged. When the potential difference betweeen the plates of the condenser reaches a certain critical value determined by the length of the primary spark-gap, the diameter of the spark balls, etc., a discharge takes place which oscillates through the primary of the oscillation transformer and back and forth across the primary spark-gap. The frequency of the current depends entirely, as shown before, on the constants of the circuit. On first thought, one would think that the condenser would discharge through the closed circuit in the transformer secondary rather than jump the air-gap, but a little consideration of the matter will show that the inductance of this circuit to electric oscillations of this nature is so great that no discharge can take place. Another matter that might be touched on here is the resistance of the spark-gap. Before any discharge has passed and under normal conditions the resistance of the spark-gap is very great: the voltage required to break down one centimeter of air being about 10,000. After the initial discharge has passed and the air becomes heated and ionized the resistance may drop as low as two or three ohms. This fact plays an important part in the damping of the oscillation trains.

The discharge from the condenser which oscillates through the primary of the oscillation transformer sets up a rapidly alternating magnetic field, which being linked with the secondary induces an electromotive force in it. The law for the induction in this case is not nearly as simple as in the case of the ordinary transformer, the capacities of the circuits playing an important part. If the capacity of the circuits is below a certain critical value, the induction is in the ratio of the capacities of the circuits, while if greater the induction depends on the relation between the number of turns in the primary and secondary. The formulæ for calculating the voltage difference across the secondary in either case are extremely complex, involving the damping factor, the capacities of the circuits, and other constants. Drude and Bjerknes have treated the subject of the oscillation transformer analytically in an admirable manner.

The frequency of the spark in the large spark-gap is not a simple one but consists of several, one being the natural period of vibration of the secondary and one a forced vibration of the secondary, due to the fact that the primary and secondary are never exactly in tune. There is also a certain small current of a high frequency, due entirely to the constants of the spark balls and connectors, which act as a condenser.

The Tesla coil in its present form is still very crude leaving much to be improved upon and wished for. The problem that presents itself in the construction of Tesla coils is practically the same one that presents itself in selective wireless telegraphy, so that the solution of the one will solve the other.

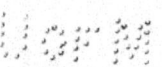

CHAPTER IX

USES OF THE COIL

THE Tesla coil readily lends itself to a great number of experiments, some interesting in their effect, others useful in scientific research. Waves for wireless messages may be sent out into space, X-Ray tubes excited, Geissler tubes illuminated, beautiful brush effects shown, and a great number of other things done.

The high potential current obtained from this coil possesses certain interesting properties due to its high frequency that are not possessed by either the Ruhmkorff induction coil or a static machine.

These properties are best seen in the beautiful brush effects, which may be obtained even with the coil described in the Appendix. All of these experiments on the brush discharge should be performed in the dark, as they then show to the best advantage.

These effects, besides affording a pleasing sight, are of great scientific value. It is a known fact that the phenomenon is due to the agitation of the molecules near the terminal, and it is thought, since the brush is hot, that some heat must be developed by the impact of the molecules against the terminal or each other. A little consideration of the matter

leads us to the conclusion that if we could but reach suffi-ciently high enough frequencies, we could produce a brush which would give intense light and heat. But this is stray-ing too much from the practical nature of this book; the only reason for putting it in being, that it might suggest to the amateur new lines for experiment.

The following experiments on the brush discharge have been taken from Nikola Tesla's "Experiments with Alternate Currents of High Potential and High Fre-quency."

There is practically nothing except some wire and a few supports required for all these experiments on the brush discharge, and thus they can be performed by every one. In the first experiment two insulated wires about 10 feet long are stretched across the room. They are supported at dis-tances of a foot and a half by insulating cords. One of the wires is attached to each terminal of the coil. When the coil is put in action the wires are seen to be strongly illumi-nated by the streams issuing abundantly from their whole surface. (This experiment must be shown in the dark, of course.) The cotton covering on the wire, although it may be very thick, does not affect the result.

To produce the best effect the primary gap and the length of the wires must be carefully adjusted. It is best to take the wires at the start very long and then adjust them by cutting off first long pieces and then shorter and shorter lengths until the correct length is reached. When this adjustment has been obtained and the wires are fed by either the 12″ or 7″ coil, the light from them will be sufficient to distinguish objects in the troom.

Another way of easily exhibiting the brush effect is by attaching the end of 10 or 20 feet of No. 36 insulated copper wire to the one terminal of the coil and the opposite end to an insulating support, leaving the wire hanging clear. Upon touching the remaining terminal with a bit of metal held in the hand, the wire will break forth into numberless streams or threads of light palpitating in unison with the discharge of the condenser.

The luminous intensity of the streams can be considerably increased by focusing them upon a small surface. This is illustrated by the following experiment. To one of the terminals of the coil a wire bent into a circle about one foot in diameter is attached and to the other terminal a small brass sphere. The centre of the sphere should be in a line at right angles to the plane of the circle at its centre. When the discharge is set up, a luminous hollow cone is formed, and in the dark one half of the brass sphere is seen strongly illuminated. To get the best results possible with this experiment, the area of the sphere should be equal to the area of the wire.

Another way in which the luminous effect of the discharge may be shown is as follows: two circles of rather stout wire, one being about 32″ in diameter and the other 12″, are formed, and to each of the terminals of the coil one of the circles is attached. The two circles must be concentric and in the same plane. When the coil is turned on the whole space between the wires is uniformly filled with streams. The intensity of the streams forming this luminous disc is such that objects in the room can be plainly distinguished even though at a considerable distance.

By this time the experimenter will realize that to pass ordinary luminous discharges through gases, no particular degree of exhaustion is necessary, but a very high frequency is essential, and of course a fairly high potential. This shows us that the attempts to produce light by the agitation of the molecules or atoms of a gas need not be limited to the vacuum tube, but the time is to be looked forward to, and that in the near future, when light will be produced without the use of any vessel whatever and with air at ordinary pressure. When light is obtained this way, there will be no chemical process nor consumption of material, but merely a transfer of energy, and the probability is that such a light would have an efficiency far exceeding that of even the best of the present incandescent lights, which waste so much of the energy in heat.

The Geissler effect can be readily shown by using only a burnt-out incandescent globe, in which the vacuum has not been destroyed. To show it the bulb should be suspended from an insulating knob, so that the discharge will pass through the centre of the bulb. The room should be dark in order to see the changes that take place in the globe. On putting the coil in action the bulb lights up with a gently pulsating, delicate purple hue. This color in a few minutes generally turns to a lovely pale green, and then sometimes, but rarely, this changes to a delicate white light. The intensity of the light is not very great, but the delicateness of the colors is something to be admired. The discharge must not be continued too long through the tube as it is liable to pierce the glass. A peculiar thing always happens if a bulb with a good filament is used, namely, that in a few minutes the filament is found to be completely shattered. With the 12" coil it is only necessary to hold the globe in the hand, without any connections to the coil, for it to light up. By moving the bulb around in the vicinity of the spark, various changes in the intensity of the light will be seen.

If the amateur is fortunate enough to possess some Geissler tubes, these may be lit up by connecting them, in series with an adjustable spark-gap, to the coil. In the case of the large coil tubes will light up when merely brought into the vicinity of the discharge.

The best X-ray tube to use with this Tesla coil is what is known as a double-focus tube, although any other tube may be used with not quite as good results. As the terminals of the X-ray tube are alternately cathode and anode, when an alternating current is used the pulsating or rather varying light, if X-rays may be called light, would be objectionable, but due to the high frequency of the current from this coil the pulsations are not noticeable. When the double-focus tube is used the rays are reflected first from one reflector and then from the other in rapid succession, and thus double the space is filled with the rays than if a single-focus tube had been used.

If the tube is a small one and is used on either the 7″ or 12″ coils, it should be connected in series with an adjustable spark-gap, to prevent any injury to it. On starting the coil this spark-gap should be open as far as possible and then gradually closed until the best result is obtained. The tube should be felt now and then to see if it has become hot, and if it becomes too hot the coil should be stopped and the spark-gap lengthened.

For those amateurs who may desire to use this coil for wireless telegraphy, the authors merely state that it is suitable, with a few minor changes, in most of the systems using a step-up transformer. For detail information they are referred to a large number of admirable books, especially devoted to this subject, written by very eminent men in this field.

But after all the great field of experimental investigation opens itself to the possessors of one of these coils when they enter the realm first mathematically investigated by Maxwell; that is, the field in which Hertz made himself famous, the field of electric waves. The subject itself is still in its infancy and great results are expected when the laws of electromagnetic disturbances are verified in a convincing manner.

It is not the easiest subject to experiment in by any means, on the other hand it is the most difficult.

In order to carry out any of the experiments at all and get satisfactory results a large room free from all metallic objects and electric wiring is required. A good place is in the loft of a large barn, provided no better place can be had. It was in a large barn that the authors verified some of Hertz's experiments on electric waves. In dealing with electric waves there are two disturbances to be taken into account, the electromagnetic and the electrostatic. There are also two classes of waves, the electromagnetic waves and the stationary waves on wires. A method will be given for obtaining both the electromagnetic and the stationary waves, which is all that can consistently be brought within the scope of this work.

D. E. Jones has made a translation from the German of the original papers of Heinrich Hertz, dealing with the experiments and results which have made his name famous.

The radiator or oscillator used by Hertz consists of two metallic plates, having attached to them short rods ending in knobs a short distance apart. These knobs are connected to the secondary of the coil. Hence, as the secondary electromotive force accumulates, the plates are brought to a difference of potential and lines of electrostatic displacement stretch out from the positive side of the oscillator to corresponding points on the negative. We have thus a strong electrostatic displacement created along the lines of force. When the potential difference reaches a certain point depending on the length of the spark-gap, the air insulation breaks down and a current flows from the one plate to the other across the spark-gap. It is merely the discharge of a condenser, for the two plates on the oscillator form a condenser, with air as a dielectric. This creates in the space around a magnetic flux, the direction of which is everywhere normal to the direction of the electric displacement. The electrostatic energy is thus transformed into electrokinetic energy.

If the electric oscillation is started sufficiently sudden some of the energy is thrown off as a displacement wave. As the coil is continually in operation we have groups of intermittent oscillations and therefore trains of electric waves thrown off which spread out through the dielectric. In this way the electromagnetic waves are set up in the air.

In order to have the breaking down of the air sufficiently sharp to obtain the oscillations, at least three things are necessary: the spark-ball surfaces must be bright and clean, no ultra-violet light must fall on the balls, and the balls must be a certain distance apart, best determined by experience.

The form of resonator devised by Hertz consists merely of a nearly closed ring or rectangle of wire the ends of which end in metallic balls placed very close together. The one ball is adjustable by the use of a micrometer screw. There have been many modifications of this resonator, but the one the authors found most satisfactory was where a neon or carbon dioxide Geissler tube of the spectrum variety was connected directly across the spark-gap. Instead of getting a spark then the tube becomes illuminated.

By using this resonator and the before-mentioned oscillator, the magnitude of the electric displacements in the space surrounding the oscillator can be mapped out.

Hertz's great work was in setting up stationary waves in air. This is accomplished by having a large metal plate set up in the room to act as a reflector. The coil with its radiator is set up in front of it, at some distance from it, so that the plane of the plates on the oscillator is parallel to the reflector. By holding the resonator parallel to the metallic reflector the nodes and antinodes can be easily traced out.

The size of the plates on the radiator used by Hertz in this experiment was 16″ square. The resonator was a ring of about No. 6 copper wire, 14″ in diameter.

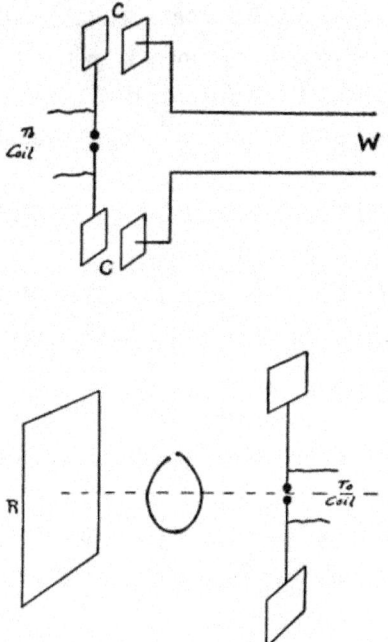

FIG. 37. — WAVES ON WIRES.

To set up stationary waves on wires the same radiator is used. Two similar plates to those on the oscillator are gotten and mounted a short distance from the plates of the radiator. The plates are all parallel and have two parallel wires led from them across the room as shown in the figure.

Thus the wires are electrostatically connected to the Hertz radiator and the plates rapidly alternate in potential, and this applies to the ends of the wires an alternating electromotive force. This EMF creates electric waves of potential which travel along the wires with the velocity of light. If the length of the wire is suitably adjusted, stationary waves of electric current and potential are set up by the interference of the direct and reflected waves.

To try the experiments on reflection, refraction polarization and interference of electric waves, a small oscillator consisting of two brass rods ending in balls and mounted in a zinc box of parabolic form is required. The resonator must be very sensitive, for otherwise no results will be obtained. The prisms and lenses used are moulded from either paraffine or pitch.

As before stated the authors said that they would merely outline the experiments possible with the Tesla coil, and for the detailed information the translations of Hertz's original works must be consulted, unless the amateur has sufficient knowledge of German to read the originals, which will of course be far more satisfactory.

There are several experiments which can be readily performed with this apparatus and which would appeal to those having no knowledge of electric phenomena. These experiments have but little practical value, the only reason for citing them being that they may furnish entertainment to some friend who should chance into his shop. The writers first saw them performed on the vaudeville stage with an Oudin resonator which is far inferior to the Tesla coil.

To begin the performance the operator should make a few general statements as to the voltage required to leap different air-gaps, such as 30,000 for 1″, about 55,000 for 2″, etc. Then show the discharge across the 6″ or 12″ gap and let the spectator imagine what voltage that represents. After he has become somewhat impressed with the intensity of the discharge, the operator can approach one of the oscillators and allow the spark to play on his bare hand. To do this without severe burning of the hand, he should keep his hand in constant motion to prevent the spark from playing on one point. There is only a very slight sensation felt when this current is traversing your body and no injurious effects result so far as the authors are aware. The effect is best shown in the dark.

The next experiment is to grasp one oscillator with the hand when the coil is in operation and to have an assistant touch the operator's bare elbow with a cotton cloth dipped in alcohol. The handkerchief immediately bursts in flame. To get the result without any uncertainty the cloth is wrapped rather tightly around the assistant's hand. An amusing modification of the same experiment is to touch the cloth to the hair of the operator, showing that the hair is not ignited.

If any of the audience are inclined to smoke, a suggestion as to lighting cigars on a windy night without the annoyance of matches blowing out will be greatly appreciated. The operator has merely to bring a piece of metal, such as a nail, held in the hand to within $\frac{1}{2}''$ of one of the spark terminals and light the cigar from the spark.

A gas flame can be lit with the bare fingers by grasping one oscillator with the bare hand and approaching the burner with the finger of the other hand so that a spark will jump to the metal tip of the burner. The writers have made this more spectacular by letting the spark jump from the tip of the tongue.

In these experiments an assistant should adjust the spark-gap so that no more current passes than is necessary. This is to prevent the spark from burning the operator.

To convince the audience of the tremendous voltage passing through the operator's body he has merely to bring one hand up to a lighted incandescent globe, while he grasps one terminal of the coil with the other hand. On the near approach of the hand the filament will violently vibrate and then shatter, blackening the bulb and of course extinguishing the lamp.

Lighting Geissler tubes held in the hand or even in the mouth by approaching them to the oscillators is an experiment that never fails to bring forth the applause of those present.

Perhaps the most spectacular experiment, one which is unaccountable for by the every-day electrician who does house wiring and has never been brought in touch with high-frequency currents is the lighting of an ordinary incandescent lamp with the current traversing the operator's body.

Before performing this experiment some few remarks on the quantity of current necessary to bring the filament to full brightness on the 110-volt circuits should be made, if those present are ignorant of electrical matters. They can thus see that the energy required to light an ordinary 16 C. P. lamp is equivalent to 55 watts, and that this amount is therefore taken through the operator's body. At 110 volts this means approximately $\frac{1}{2}$ ampere, while according to the best authorities $\frac{1}{10}$ of an ampere is fatal to the average human. The reason why this amount of energy can be taken with impunity is not definitely known, but it is thought to be due to the fact that the high-frequency current does not penetrate into the interior of a solid conductor, but follows the surface. This is known as the skin effect.

The operator, to get the best results, should stand on an insulated stool and grasp one terminal of the coil with one hand, and approach with a piece of metal held tightly in the hand or mouth one lead of a lamp, the other lead of which has been previously grounded. The lamp will come up to bright red and if an assistant adjusts the primary spark-gap to its best working distance, the lamp may be brought up to full brightness.

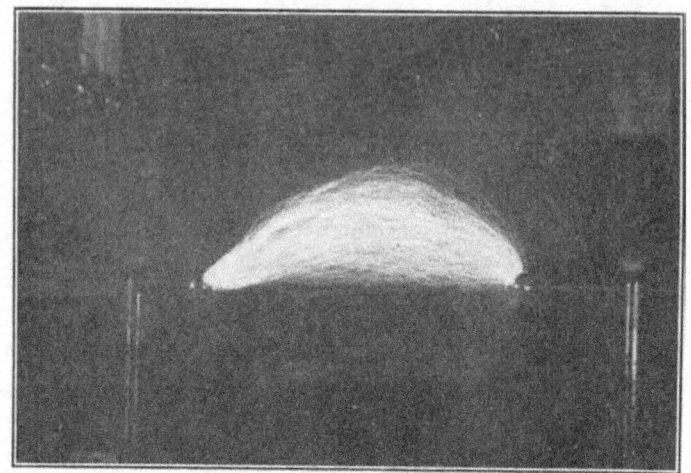

PLATE VI. — Discharge from the 12″ Coil.

PLATE VII. — The 7″ Standard Apparatus.

CHAPTER X

DIMENSIONS OF 7" STANDARD COIL

FOR those amateurs who, having read the previous chapters, think that an apparatus giving a twelve-inch spark is too large for their limited uses, this chapter has been added.

This coil is by no means to be thought of as a toy, for the authors themselves used the very apparatus described in this chapter in carrying out their first experiments with the X-ray and Geissler tubes. Wireless messages were also sent successfully over a distance of three miles in wet weather. This was the greatest available distance over which the authors could try the coil, so that three miles should not be considered the maximum transmitting distance. In clear weather messages could easily be sent a distance of about fifty miles, provided your antennæ is well insulated from grounds.

Because this apparatus is not as powerful as the other, does not mean that any less care should be taken as regards insulation and mechanical construction, for it depends entirely on this whether the coil builder is to get a thin, intermittent spark or a fat crackling one. The only difference between this apparatus and the larger one, besides that of

97

size, is in the construction of the transformer and condenser
and then they are only trivial.

The core of the transformer is 2⅝″ in diameter and is
built up of pieces of No. 32 B. & S. gauge iron wire 13″
long after the manner described in Chapter II. The same
care should be taken in annealing and insulating the iron
wires as was done before.

The primary is wound in two sections adjacent to each
other, as seen in Fig. 38. Each section is wound towards

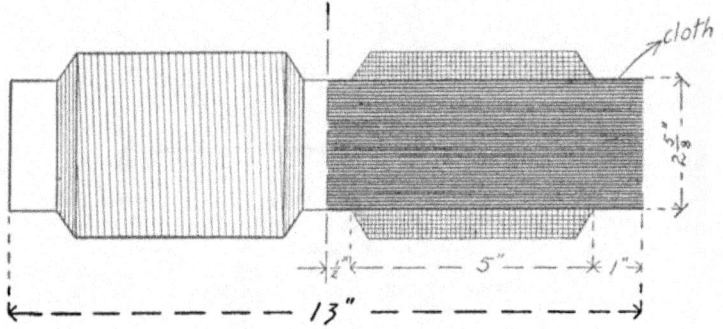

FIG 38. — PRIMARY AND CORE OF TRANSFORMER OF 7″ COIL.

the centre, starting 1″ from the ends of the core, for a distance
of 5″. There are six layers of No. 16 B. & S. gauge double
cotton covered copper wire in each section. Each layer is
thoroughly shellacked when put on and the terminal wires
are held by the same method as described in Chapter II.
At least two feet should be left for bringing out the terminals
to the binding-posts.

The secondary is wound in two sections. No. 32 B. & S.
gauge double cotton covered copper wire is used. The tube

on which it is wound has an internal diameter of 4" and the thickness of the wall is $\frac{1}{16}$". It is 11" long and made from the best vulcanized fibre. The bobbin heads are cut out of $\frac{1}{4}$" sheet fibre. They are 6" in diameter and have a hole $4\frac{1}{8}$" in diameter cut out of the centre. Four of these are

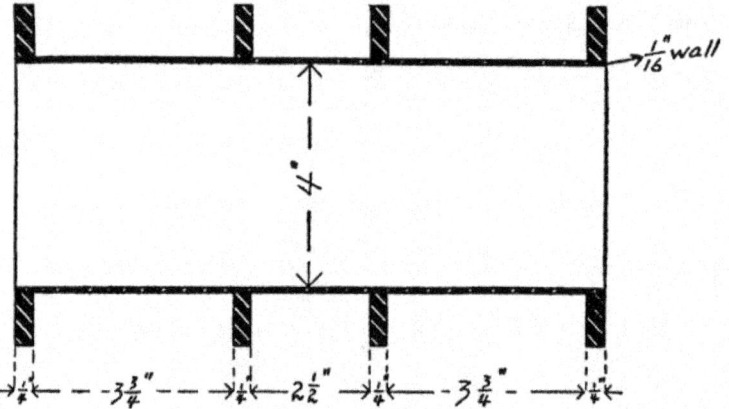

FIG. 39. — SECONDARY BOBBIN OF TRANSFORMER OF 7" COIL.

required. They are slipped on the tube to the positions shown in the figure. The distance between the bobbin heads of each section is $3\frac{3}{4}$", and the distance between the two sections $2\frac{1}{2}$". The bobbin heads can be held in place in the same manner as described in Chapter II by cutting the rings out of a very thin fibre tube, or in this case it will be sufficient to wrap some heavy brown paper several times around the tube between the bobbin heads. After winding a layer on the secondary, shellac it and wrap it with one turn of paper. In this way build up the secondary to within $\frac{1}{8}$" of the bobbin heads. When the last layer is put on it is wrapped with several turns of paper which is shellacked in place. This completes the construction of the transformer. When finished it should be left for some time in a warm dry place as behind the stove, to thoroughly dry the shellac. The reason for this is that green shellac is a fairly good conductor.

For the condenser, twenty-eight sheets of brass 6″ x 10″ are required. No. 32 or 34 soft sheet brass is used. Each sheet has a lug $1\frac{1}{8}$″ long and $1\frac{1}{2}$″ wide either cut directly on it or soldered on in the upper corner. Whether they are cut directly on the sheet or are soldered on will depend on the width of the brass sheet used. With 12″ brass they are cut on the sheet or soldered on with 8″ brass. A $\frac{1}{8}$″ lip is bent across the top of each lug. See Fig. 40. Thirty sheets of glass $\frac{1}{10}$″ thick and 7″ x 12″ are required. Any sheet of glass that has an air bubble in it should be rejected as it is liable sooner or later to give way, thus causing the reconstruction of the condenser. Wipe each sheet of glass clean before putting it in to the condenser.

The frame is constructed from well seasoned pine. The two sides are made from $\frac{1}{2}$″ x 12″ pine 4″ long. The base is made of a piece of $\frac{1}{4}$″ x 4″ pine 8″ long. The base is fastened to the sides by some flat-headed brass screws. The heads of the screws should be sunk flush with the wood. On the one side of the frame two strips $\frac{1}{4}$″ x 1″ are fastened; one at the top and the other at the bottom of the frame. See figure. The frame is now laid on a flat surface with the side on which the strips have just been fastened down.

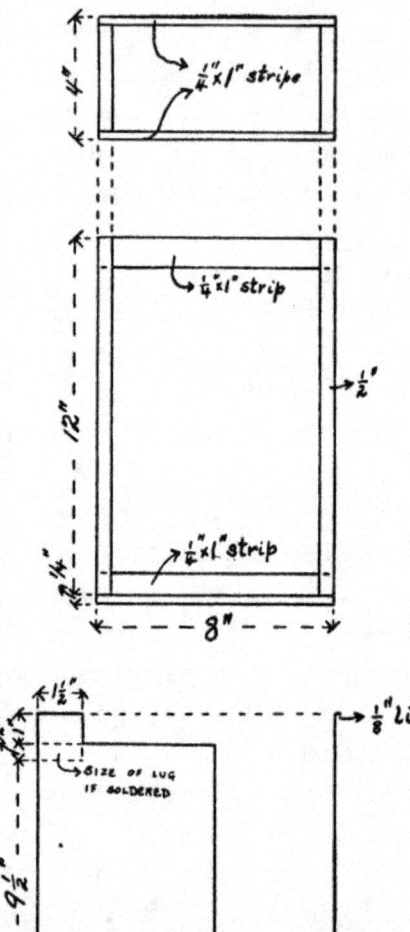

FIG. 40. — PLATE AND FRAME OF CONDENSER.

A glass plate is placed in the frame after having been wiped dry and clean so that it touches the bottom of the frame. Then a brass sheet is laid in so that there is 1″ margin of glass at the bottom and a ½″ margin on the sides. The lip should just fit against the upper edge of the glass. Without displacing the brass sheet place a sheet of glass on top of it. This is followed with a sheet of brass, but in this case the lug on the brass is brought out on the opposite side to the previous one. Continue this process until the 28 sheets of glass have been put in place. Two glass sheets are placed on top of the last brass sheet. Also remember to bring the lugs from the brass sheets out on alternate sides.

Mortices should be cut in the top and bottom of the upper ends of the sides a little deeper than the point to which the last glass sheet reaches. These are to receive two strips of pine ¼″ x 1″ similar to those on the other side. These should be screwed down so that they press firmly against the glass. A piece of paper or cloth placed between the strip and the glass will prevent the breaking of the latter.

Set the condenser upright and solder a piece of copper wire, which has already been tinned, to each of the lips in turn down the one side and another wire to all the lips on the other side. About No. 16 bare wire will do. Enough extra wire should be left to make all necessary connections.

The oscillation transformer is constructed in the same manner as the one for the 12″ coil. The circular supports for the secondary are 6″ in diameter and are turned out of ½″ material. Eight equidistant slots are cut in the per-

iphery $\frac{1}{2}"$ square. The fibre strips are $\frac{1}{2}"$ square and 11" long. A rod 11" long is turned out to the size shown in the figure 41. It is 1" in diameter and has a $\frac{1}{2}"$ shoulder turned on each end. This rod holds the two supports for the secondary apart. If the method of winding the wire in grooves is to be used, the thread should be cut on the fibre strips before mounting them on the supports.

In the original coil the wire was merely wound on three fibre supports 3" wide mounted on a hexagonal end piece. The wire was wound so that no adjacent turns were in contact and the whole was thoroughly shellacked. Although this method of winding gave good results while the coil was new, it was found after some usage that the wires became loosened, thereby reducing the effective sparking distance.

A better way however, is to use one of the methods described in Chapter IV. The wire used is No. 28 B. & S. gauge double cotton covered copper wire and was wound 18 turns to the inch.

The end pieces for the primary are cut out of $\frac{1}{2}"$ material and are 9" square. The diameter of the circle in which the dowels fit is 8". There are twenty-four $\frac{1}{4}"$ maple dowels used in all. After the secondary is wound these end pieces are screwed to the secondary frame, and the dowels slipped in place. The length of the frame over all is 12". The primary winding consists of one and a half turns of a copper ribbon $\frac{1}{2}"$ wide. The turns should be equally spaced and the winding stretch from the one end of the frame to the other. A copper wire is soldered on to each end of the

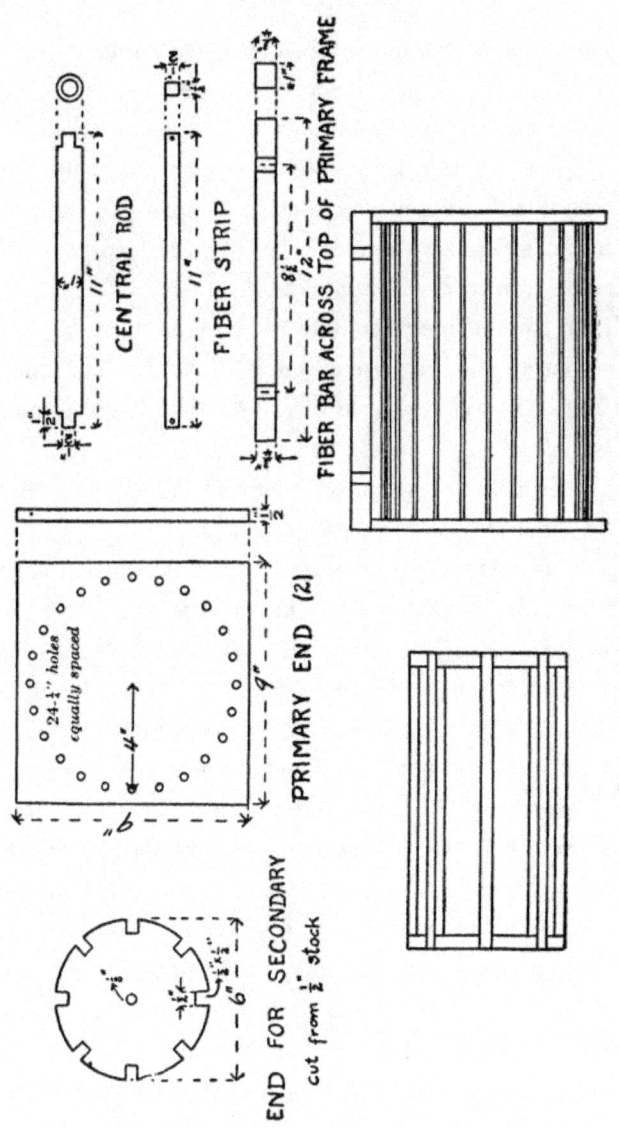

CENTRAL ROD

FIBER STRIP

FIBER BAR ACROSS TOP OF PRIMARY FRAME

PRIMARY END (2)

24-¼" holes equally spaced

END FOR SECONDARY
cut from ½" stock

FIG. 41 — OSCILLATION TRANSFORMER OF 7" APPARATUS.

primary band for making the connections. A hard rubber strip 1″ x ¾″, 12″ long, is screwed across the top of the completed oscillation transformer. Two holes are drilled in it 8½″ apart. Into these two bushings, similar to those used in the 12″ coil and described in Chapter IV, are fitted. When the secondary terminals are soldered to them the oscillation transformer is complete.

All the parts of the 7″ coil are mounted in one box. The dimensions of this box are given in Fig. 42. The box is built out of ¾″ well seasoned oak. All the directions given in Chapter VI apply to the construction of this box. The cover is divided into two halves, one carrying the interrupter and the other the discharge oscillators.

The connections from the interrupter to the condenser are made through the hinges so that the cover may be swung back without disturbing the connections. A partition is put in between the transformer and the condenser. It has a number of holes drilled in it to allow of the free circulation of the oil. Suitable handles are put at each end of the box for carrying it.

The transformer is now set on end in the smaller division of the box. It is held in place by two yoke-shaped, wooden supports fastened to the inside of the box and encircling the core between the two sections of the secondary. The primary terminals are brought to four heavy binding-posts at the upper end of the box. They should be soldered on in the same order as for the transformer on the large coil, that is, so that shorting the two middle posts puts the sections in series and shorting the two outer pairs gives a parallel connection.

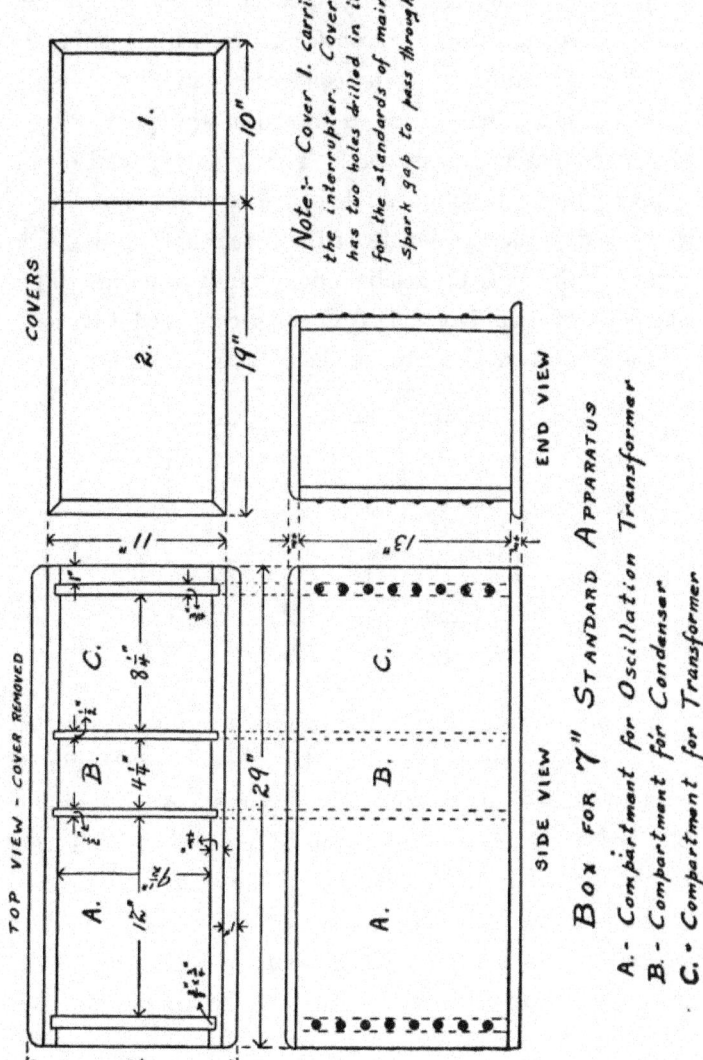

COVERS

1.

2.

10"

19"

Note:- Cover 1. carries the interrupter. Cover 2 has two holes drilled in it for the standards of main spark gap to pass through.

11"

END VIEW

13"

TOP VIEW - COVER REMOVED

C.

8¼"

B.

4¼"

A.

12"

29"

13½"

SIDE VIEW

C.

B.

A.

Box for 7" Standard Apparatus

A.- Compartment for Oscillation Transformer
B.- Compartment for Condenser
C.- Compartment for Transformer

Fig. 42. — Box for 7" Apparatus.

The condenser and oscillation transformer are now put in place, the condenser being between the two. The secondary terminals from the transformer are led in glass tubes, suitably bent, directly to the condenser. From one side of the condenser a wire is led to an end of the primary band on the oscillation transformer. The remaining end of the copper band and the other side of the condenser are directly connected to the two hinges of the cover carrying the interrupter. All connections should be carefully soldered. They should be of about No. 20 B. & S. gauge copper wire, enclosed in glass tubes and kept under the oil as much as possible.

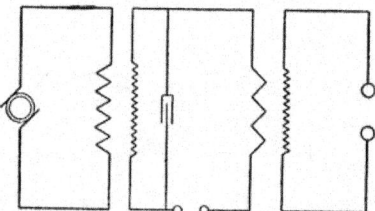

FIG. 43. — WIRING DIAGRAM.

Any one of the forms of interrupters described in Chapter V can be used with the coil; the coil in question being equipped with the motor interrupter. The connections between the primary spark-gap and hinges can be run in glass tubes lying in grooves cut in the under side of the cover. A piece of ¼" hard rubber sheet should be screwed over the grooves wherever there is any danger of shorting to the core of the transformer or primary terminals. The connections are shown in the wiring diagram.

The oscillators consist of two brass balls $\frac{3}{4}''$ in diameter screwed on the end of two $\frac{3}{16}''$ brass rods $7''$ long, which are to slide easily in two holes drilled $\frac{1}{2}''$ from the top of the standards, through both the fibre and the rod. A set screw at the top of each standard will be of convenience in clamping the rods in any desired position.

The standards are constructed as follows. Two fibre or hard rubber bushings $2''$ in diameter and $1\frac{1}{2}''$ in length and having a flange $\frac{1}{4}''$ thick and $2\frac{1}{2}''$ in diameter turned on one end are set in two holes cut in the cover directly above the holes in the brass bushings on the oscillation transformer. A $\frac{3}{4}''$ hole is drilled through the centre of each bushing. Two $\frac{3}{8}''$ brass rods $8''$ long are enclosed in fibre tubes $\frac{3}{4}''$ in outside diameter and $7\frac{1}{2}''$ long. The tubes should fit the rods tightly. The ends of the brass rods project from the fibre and should be slightly tapered to fit the bushings on the oscillation transformer.

In order that the discharge gap may be adjusted while the coil is in operation, two vulcanite handles $\frac{3}{4}''$ in diameter are screwed on the ends of the rods, carrying the oscillators, for about $1\frac{1}{2}''$.

The standards are now slid through the bushings in the cover until they make good contact with the bushings on the oscillation transformer. When the coil is now connected up to the alternating current mains, it will break forth in a beautiful $7''$ discharge.

If everything is not as it ought to be, the trouble may be found in the manner described in Chapter VII.

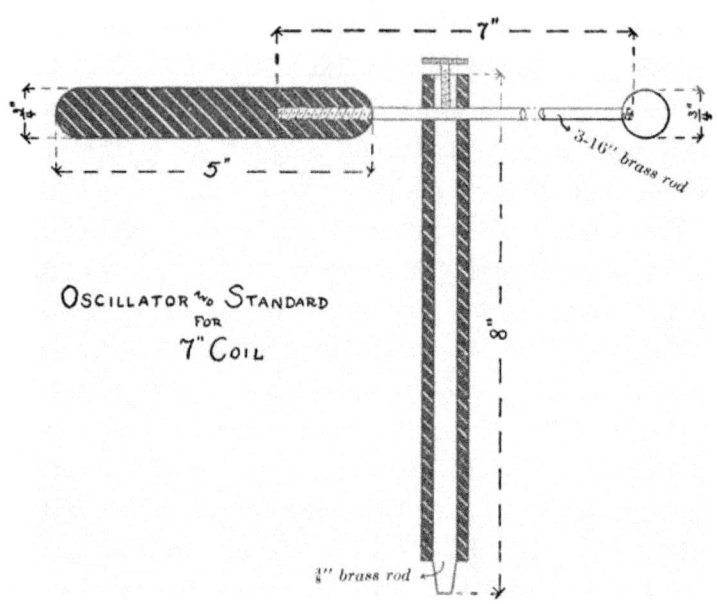

OSCILLATOR AND STANDARD
FOR
7" COIL

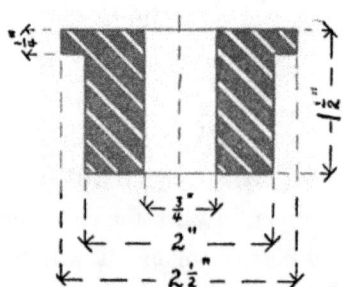

BUSHING FOR 7" COIL

FIG. 44. — OSCILLATORS AND STANDARDS FOR 7" APPARATUS.

APPENDIX

For those of our readers who have limited means at their disposal, and who desire to carry on some of the many experiments possible with high-tension currents, this Appendix has been added. Besides many are not situated in cities, where an alternating-current lighting supply is available, but who possess an ordinary induction coil, giving a two or three inch spark, which they may substitute for the transformer to be described in the present article.

This coil is not oil immersed, hence no boxes will be required, as it is simply mounted on a base in a place free from dust and moisture. A large amount of the precautions regarding insulation and other things can be dispensed with, thereby reducing the cost of the materials to within the reach of almost every one. While speaking of cost, let us state that to purchase a coil giving a 12″ spark from the regular dealers would mean an outlay of about $300, while the 7″ coil in a single box is worth $165. The cost of construction by the amateur, not considering his time, should not exceed $50 for the 12″ coil and $25 for the 7″ coil.

This piece of apparatus giving about a 3″ spark should not exceed $10 to build at home. It is large enough for most of the experiments on Roentgen and Geisler tubes and for wireless work over short distances.

The above sums include the simple interrupter. The others will bring the price up in proportion.

The high-frequency coil is made as follows: Cut out two end pieces of 1″ wood 10″ square and describe on each one two concentric circles, having diameters of 9 and 7 inches respectively. On these circles bore a number of ¼″ holes 1″ apart as in the figure. Next procure from a planing mill about twenty ¼″ dowels. These are made of hard wood and come 36″ long. Cut each dowel into 12″ lengths and fit one in each of the holes on the smaller circle of one of the boards. When they are all in place the other board is put on the other end of the dowels. The outer circle of holes is left empty until the secondary is wound.

The secondary winding consists of one layer of No. 32 B. & S. gauge double cotton covered copper wire. Begin the winding about ½″ from the ends. Shellac the wire with several coats of the best orange shellac when the winding is finished.

The dowels for the primary are next put in place by pushing them through the holes from one end. If they fit too tightly the holes may be reamed out. Next six turns of No. 18 bare wire are wound on the outer dowels, each turn being over an inch from the one next to it.

The whole coil is then mounted on a base. The ends of the primary are connected to two binding-posts mounted on a piece of hard rubber. Two oscillators with standards are provided for the terminals of the secondary. This completes the high-tension coil.

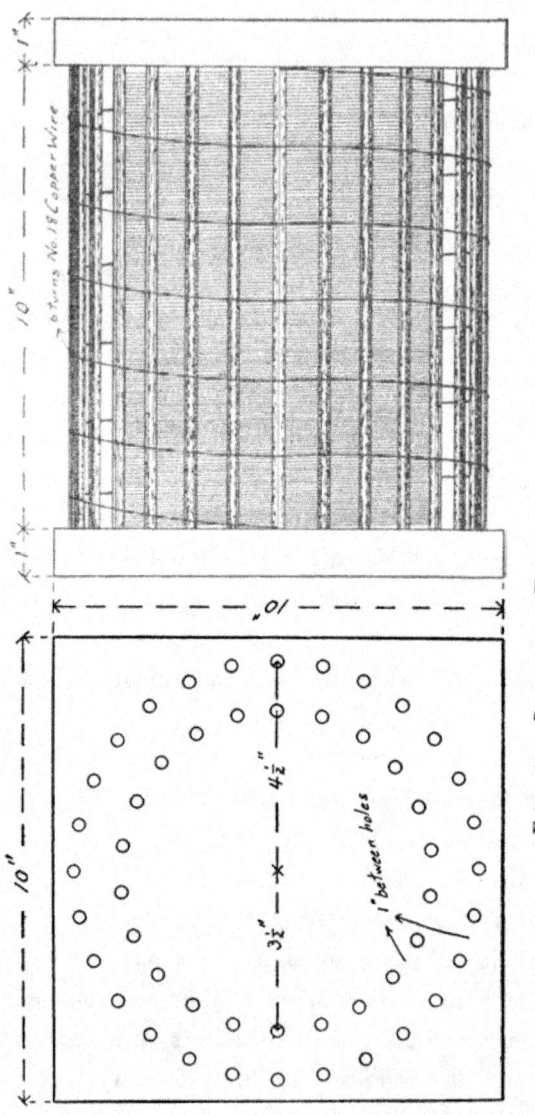

FIG. 45. — OSCILLATION TRANSFORMER OF SMALL COIL.

The condenser consists of fifteen sheets of window glass 10″ x 12″, with a piece of tin foil 8″ x 10″ between each sheet of glass. The method of arranging this condenser is as follows: Lay a glass plate on a smooth table and give it a coat of shellac. While still wet place a sheet of tin foil on top of it, leaving an inch margin of glass all around. On one corner lay a strip of tin foil projecting an inch beyond the glass. On top of this lay a second sheet of glass and another sheet of tin foil, only the strip in this case is brought out on the opposite side. Continue this until the fifteen sheets of glass are used up. This will give seven sheets of tin foil, with the strips coming out on the one side and seven projecting on the other side. The strips may be fused together with a hot iron and a copper wire soldered on. The whole condenser is bound together with insulating tape and is best mounted in a box. This completes the condenser.

The transformer for use with the 100–110 or 50–55 volt alternating-current circuits is the next piece of apparatus to construct. It is essentially the same as the two transformers already described. The core consists of a bundle of No. 20-22 iron wires, well annealed. The diameter is 1½″ and when formed after the method described in Chapter II is wrapped with insulating tape. The primary is wound in two sections one above the other. See Fig. 1 for the method of fastening the layers. Each section consists of one layer of No. 16 B. & S. gauge double cotton covered copper wire. After the primary is wound wrap on several layers of paper well shellacked until the diameter is built up to 2¼″.

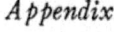

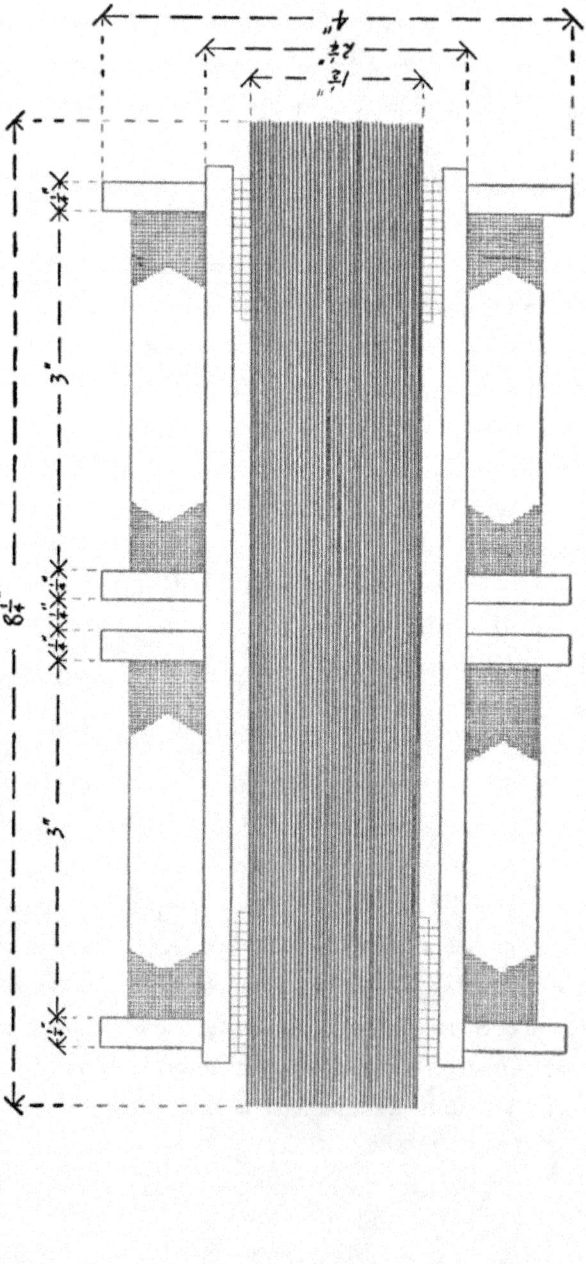

FIG. 46. — COMPLETED TRANSFORMER OF SMALL COIL.

The secondary winding of this transformer consists of two sections of No. 32 B. & S. gauge double cotton covered copper wire. First saw out of $\frac{1}{4}''$ stock four circular pieces of wood, 4″ in diameter and having a $2\frac{1}{4}''$ hole in the centre. Slip these on the primary to the positions shown in the figure. The two end ones are $\frac{1}{2}''$ from the ends of the core and the middle ones are $\frac{1}{4}''$ apart.

Wind the wire of the secondary on the two spools just formed until the diameter is $3\frac{1}{2}''$. Thoroughly shellac each layer and wrap a piece of paper on before beginning the next. The whole coil is mounted on a suitable base, the primary terminals being connected to binding-posts.

If the transformer is to be operated on the 100–110 volt current, the two sections of the primary are connected in series. If, on the other hand, it is to be used on the 50–55 volt current the sections are joined in parallel. It is well, however, in either case to bring the primary terminals out to four separate binding-posts. Then the desired connections may be readily made for either series or parallel. Always be certain, though, that the current will traverse the windings in the same direction.

In order to set up the high oscillations we must introduce a spark-gap in series with the secondary of the transformer and the high-tension coil. The method of making this primary spark-gap is given as follows: Procure two pieces of vulcanized fibre rod $\frac{3}{4}''$ in diameter 4 inches long. Drill a $\frac{1}{4}''$ hole in each $\frac{1}{2}''$ from one end. Next bore two $\frac{3}{4}''$ holes 6″ apart in the base of the transformer as shown in Fig. 47.

Drive the fibre supports into these holes with the holes in the fibre in line.

The spark-gap is made of two ¼″ brass rods 6″ long with fibre tube 2″ long slipped over the end to act as an insulating handle. One lead of the secondary of the transformer goes directly to one rod, the other goes to the primary of the high-tension coil. The return wire from the primary of the high-tension coil is soldered to the other side of the spark-gap. The diagram shows how the condenser is connected and also the connections just described.

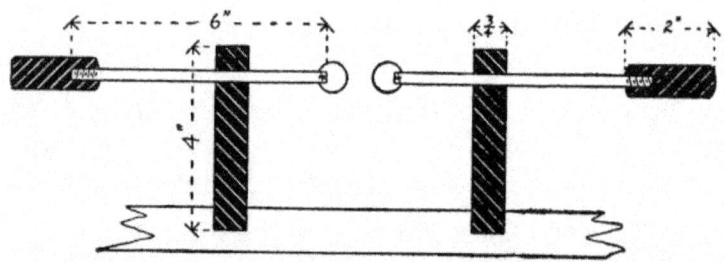

Fig. 47. — Primary Spark-gap.

There is no interrupter used with this apparatus so that care must be taken that the spark is long enough to prevent arcing.

Those possessing a suitable induction coil and who wish to substitute this for the transformer and primary spark-gap may do so by changing one connection. Disconnect one terminal of the secondary from the discharger and connect the secondary terminal to a binding-post suitably insulated by hard rubber. One terminal of the primary of the high-

tension coil is connected to the spark-gap instead of the secondary of the induction coil.   The other terminal of the high-
tension coils primary is connected to the new binding-post.
A glance at the figure will make this plain and also the method

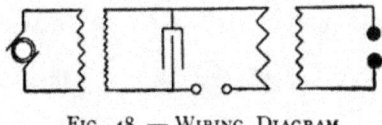

FIG. 48. — WIRING DIAGRAM.

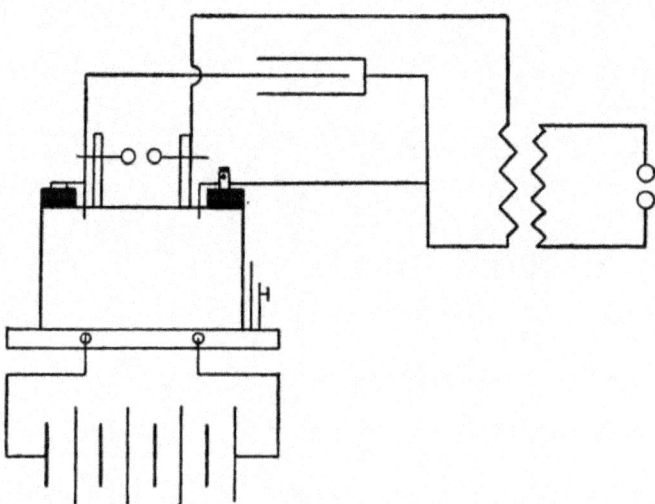

FIG. 49. — WIRING DIAGRAM.

of connecting up the condenser.   When making connections
between the various parts of the apparatus it is well to enclose
the wires in glass tubes and to keep them back out of the way.
The operator will soon find that ordinary insulation is of

no value whatever in dealing with these high-tension currents, so that all terminals must be kept apart a distance greater than that of the high-tenson discharge gap. If this precaution is not observed you will have some very beautiful brush discharges all along the conductors that are in too close proximity.

LIST OF WORKS

ON

ELECTRICAL SCIENCE

PUBLISHED AND FOR SALE BY

D. VAN NOSTRAND COMPANY,

23 Murray and 27 Warren Streets, New York.

ABBOTT, A. V. The Electrical Transmission of Energy. A Manual for the Design of Electrical Circuits. *Fifth Edition, enlarged and rewritten.* With many Diagrams, Engravings and Folding Plates. 8vo., cloth, 675 pp. . . . . .Net, $5.00

ADDYMAN, F. T. Practical X-Ray Work. Illustrated. 8vo., cloth, 200 pp. . . . . .Net, $4.00

ALEXANDER, J. H. Elementary Electrical Engineering in Theory and Practice. A class-book for junior and senior students and working electricians. Illustrated. 12mo., cloth, 208 pp. $2.00

ANDERSON, GEO. L. Handbook for the Use of Electricians in the operation and care of Electrical Machinery and Apparatus of the United States Seacoast Defenses. Prepared under the direction of Lieut.-General Commanding the Army. Illustrated 8vo., cloth, 161 pp. $3.00

ARNOLD, E. Armature Windings of Direct-Current Dynamos. Extension and Application of a general Winding Rule. Translated from the original German by Francis B. DeGress. Illustrated. 8vo. cloth, 124 pp. $2.00

121

ASHE, S. W. Electricity Experimentally and Practically Applied. 422 illustrations. 12mo., cloth, 375 pp...............Net, $2.00

ASHE, S. W., and KEILEY, J. D. Electric Railways Theoretically and Practically Treated. Illustrated. 12mo., cloth.
Vol. I. Rolling Stock. *Second Edition.* 285 pp..........Net, $2.50
Vol. II. Substations and Distributing Systems. 296 pp....Net, $2.50

ATKINSON, A. A. Electrical and Magnetic Calculations. For the use of Electrical Engineers and others interested in the Theory and Application of Electricity and Magnetism. *Third Edition, revised.* Illustrated. 12mo., cloth, 310 pp. ....Net, $1.50

ATKINSON, PHILIP. The Elements of Dynamic Electricity and Magnetism. *Fourth Edition.* Illustrated. 12mo., cloth, 405 pp. .$2.00

Elements of Electric Lighting, including Electric Generation, Measurement, Storage, and Distribution. *Tenth Edition,* fully revised and new matter added. Illustrated. 12mo., cloth, 280 pp...........$1.50

Power Transmitted by Electricity and Applied by the Electric Motor, including Electric Railway Construction. Illustrated. *Fourth Edition,* fully revised and new matter added. 12mo., cloth, 241 pp....$2.00

AYRTON, HERTHA. The Electric Arc. Illustrated. 8vo., cloth, 479 pp...Net, $5.00

AYRTON, W. E. Practical Electricity. A Laboratory and Lecture Course. Illustrated. 12mo., cloth, 643 pp................$2.00

BAKER, J. T. The Telegraphic Transmission of Photographs. 63 illustrations. 12mo., cloth, 155 pp....................Net, $1.25

BEDELL, FREDERICK. Direct and Alternating Current Testing. Assisted by C. A. Pierce. Illustrated. 8vo., cloth. 250 pp., Net, $2.00

BEDELL, F. & CREHORE, ALBERT C. Alternating Currents. An analytical and graphical treatment for students and engineers. *Fifth Edition.* 112 illustrations. 8vo., cloth, 325 pp...Net, $2.50

BIGGS, C. H. W. First Principles of Electricity and Magnetism. Illustrated. 12mo., cloth, 495 pp............................$2.00

BONNEY, G. E. The Electro-Plater's Hand Book. A Manual for Amateurs and Young Students of Electro-Metallurgy. *Fourth Edition, enlarged.* 61 Illustrations. 12mo., cloth, 208 pp...........$1.20

122

BOTTONE, S. R.   Magnetos For Automobilists; How Made and How Used. A handbook of practical instruction on the manufacture and adaptation of the magneto to the needs of the motorist. *Second Edition, enlarged.* 52 illustrations. 12mo., cloth, 118 pp.......Net, $1.00

Electric Motors, How Made and How Used. Illustrated. 12mo., cloth, 168 pp...75 cents

BOWKER, WM. R.   Dynamo, Motor, and Switchboard Circuits for Electrical Engineers: a practical book dealing with the subject of Direct, Alternating, and Polyphase Currents.   *Second Edition, greatly enlarged*, 130 illustrations. 8vo., cloth, 180 pp..........Net, $2.50

CARTER, E. T.   Motive Power and Gearing for Electrical Machinery; a treatise on the theory and practice of the mechanical equipment of power stations for electric supply and for electric traction. *Second Edition, revised.* Illustrated. 8vo., cloth, 700 pp.......Net, $5.00

CHILD, CHAS. T.   The How and Why of Electricity: a book of information for non-technical readers, treating of the properties of Electricity, and how it is generated, handled, controlled, measured, and set to work. Also explaining the operation of Electrical Apparatus Illustrated. 8vo., cloth, 140 pp...........................$1.00

CLARK, D. K.   Tramways, Their Construction and Working. *Second Edition.* Illustrated. 8vo., cloth, 758 pp.................$9.00

COOPER, W. R.   Primary Batteries: their Theory, Construction, and Use 131 Illustrations. 8vo., cloth, 324 pp..................Net, $4.00

The Electrician Primers.   Being a series of helpful primers on electrical subjects, for use of students, artisans, and general readers. *Second Edition.* Illustrated. Three volumes in one. 8vo., cloth..Net, $5.00

Vol. I.—Theory...Net, $2.00

Vol. II.—Electric Traction, Lighting and Power...........Net, $3.00

Vol. III.—Telegraphy, Telephony, etc.....................Net, $2.00

CROCKER, F. B.   Electric Lighting.   A Practical Exposition of the Art for the use of Electricians, Students, and others interested in the Installation or Operation of Electric-Lighting Plants.

Vol. I.—The Generating Plant. *Seventh Edition, entirely revised.* Illustrated. 8vo., cloth, 482 pp............................$3.00

Vol. II.—Distributing System and Lamps. *Sixth Edition.* Illustrated 8vo., cloth, 505 pp.......................................$3.00

CROCKER, F. B., and ARENDT, M.   Electric Motors: Their Action, Control, and Application.   160 illustrations.   8vo., cloth, 296 pp. Net, $2.50

123

CROCKER, F. B., and WHEELER, S. S. The Management of Electrical Machinery. Being a *thoroughly revised and rewritten edition* of the authors' "Practical Management of Dynamos and Motors." *Seventh Edition.* Illustrated. 16mo., cloth, 232 pp......Net, $1.00

CUSHING, H. C., Jr. Standard Wiring for Electric Light and Power. Illustrated. 16mo., leather, 156 pp........................$1.00

DAVIES, F. H. Electric Power and Traction. Illustrated. 8vo., cloth, 293 pp. (Van Nostrand's Westminster Series.).........Net, $2.00

DAWSON, PHILIP. Electric Traction on Railways. 610 Illustrations. 8vo., half leather, 891 pp................................Net, $9.00

DEL MAR, W. A. Electric Power Conductors. 69 illustrations. 8vo., cloth, 330 pp...Net, $2.00

DIBDIN, W. J. Public Lighting by Gas and Electricity. With many Tables, Figures, and Diagrams. ·Illustrated. 8vo., cloth, 537 pp.Net, $8.00

DINGER, Lieut. H. C. Handbook for the Care and Operation of Naval Machinery. *Second Edition.* 124 Illustrations. 16mo., cloth, 302 pp..Net, $2.00

DYNAMIC ELECTRICITY: Its Modern Use and Measurement, chiefly in its application to Electric Lighting and Telegraphy, including: 1. Some Points in Electric Lighting, by Dr. John Hopkinson. 2. On the Treatment of Electricity for Commercial Purposes, by J. N. Shoolbred. 3. Electric-Light Arithmetic, by R. E. Day, M.E. *Fourth Edition.* Illustrated. 16mo., boards, 166 pp. (No. 71 Van Nostrand's Science Series.)................................50 cents

EDGCUMBE, K. Industrial Electrical Measuring Instruments. Illustrated. 8vo., cloth, 227 pp...........................Net, $2.50

ERSKINE-MURRAY, J. A Handbook of Wireless Telegraphy: Its Theory and Practice. For the use of electrical engineers, students, and operators. *Second Edition, revised and enlarged.* 180 Illustrations. 8vo., cloth, 388 pp....................................Net, $3.50

—— Wireless Telephones and How they Work. Illustrated. 16mo., cloth, 75 pp...$1.00

EWING, J. A. Magnetic Induction in Iron and other Metals. *Third Edition, revised.* Illustrated. 8vo., cloth, 393 pp.......Net, $4.00

FISHER, H. K. C., and DARBY, W. C. Students' Guide to Submarine Cable Testing. *Third Edition, new, enlarged.* Illustrated. 8vo., cloth, 326 pp...Net, $3.50

FLEMING, J. A., Prof. The Alternate-Current Transformer in Theory and Practice.

Vol. I.: The Induction of Electric Currents. *Fifth Issue.* Illustrated. 8vo., cloth, 641 pp.....................................Net, $5.00

Vol. II.: The Utilization of Induced Currents. *Third Issue.* Illustrated. 8vo., cloth, 587 pp...........................Net, $5.00

Handbook for the Electrical Laboratory and Testing Room. Two Volumes. Illustrated. 8vo., cloth, 1160 pp. Each vol.....Net, $5.00

FOSTER, H. A. With the Collaboration of Eminent Specialists. Electrical Engineers' Pocket Book. A handbook of useful data for Electricians and Electrical Engineers. With innumerable Tables, Diagrams, and Figures. The most complete book of its kind ever published, treating of the latest and best Practice in Electrical Engineering. *sixth Edition, completely revised and enlarged.* Fully Illustrated. Pocket Size. Leather. Thumb Indexed. 1636 pp........$5.00

FOWLE, F. F. The Protection of Railroads from Overhead Transmission Line Crossings. 35 illustrations. 12mo., cloth, 76 pp. Net, $1.50.

GANT, L. W. Elements of Electric Traction for Motormen and Others. Illustrated with Diagrams. 8vo., cloth, 217 pp.........Net, $2.50

GERHARDI, C. H. W. Electricity Meters; their Construction and Management. A practical manual for engineers and students. Illustrated. 8vo., cloth, 337 pp...........................Net, $4.00

GORE, GEORGE. The Art of Electrolytic Separation of Metals (Theoretical and Practical). Illustrated. 8vo., cloth, 295 pp. ....Net, $3.50

GRAY, J. Electrical Influence Machines: Their Historical Development and Modern Forms. With Instructions for making them. *Second Edition, revised and enlarged.* With 105 Figures and Diagrams. 12mo., cloth, 296 pp......................................$2.00

GROTH, L. A. Welding and Cutting Metals by Aid of Gases or Electricity. 124 illustrations. 8vo., cloth, 280 pp.....Net, $3.00

HALLER, G. F. and CUNNINGHAM, E. T. The Tesla High Frequency Coil; its construction and uses. 12mo., cloth, 56 illustrations, 130 pp..*In Press*

HAMMER, W. J. Radium, and Other Radio Active Substances; Polonium, Actinium, and Thorium. With a consideration of Phosphorescent and Fluorescent Substances, the properties and applications of Selenium, and the treatment of disease by the Ultra-Violet Light. With Engravings and Plates. 8vo., cloth, 72 pp.............$1.00

125

HARRISON, N. Electric Wiring Diagrams and Switchboards. Illustrated. 12mo., cloth, 272 pp.............................$1.50

HASKINS, C. H. The Galvanometer and its Uses. A Manual for Electricians and Students. *Fifth Edition, revised.* Illustrated. 16mo., morocco, 75 pp...$1.50

HAWKINS, C. C., and WALLIS, F. The Dynamo: Its Theory, Design, and Manufacture. *Fourth Edition, revised and enlarged.* 190 Illustrations. 8vo., cloth, 925 pp...............................$3.00

HAY, ALFRED Principles of Alternate-Current Working. *Second Edition.* Illustrated. 12mo., cloth, 390 pp........................$2.00

Alternating Currents; their theory, generation, and transformation. *Second Edition.* 191 Illustrations. 8vo., cloth, 319 pp...Net, $2.50

An Introductory Course of Conti uous-Current Engineering. Illustrated. 8vo., cloth, 327 pp............................Net, $2.50

HEAVISIDE, O. Electromagnetic Theory. Two Volumes with Many Diagrams. 8vo., cloth, 1006 pp. Each Vol............Net, $5.00

HEDGES, K. Modern Lightning Conductors. An illustrated Supplement to the Report of the Research Committee of 1905, with notes as to methods of protection and specifications. Illustrated. 8vo., cloth, 119 pp.......................................Net, $3.00

HOBART, H. M. Heavy Electrical Engineering. Illustrated. 8vo., cloth, 338 pp..Net, $4.50

—— Electricity. A text-book designed in particular for engineering students. 115 illustrations. 43 tables. 8vo., cloth, 266 pp., Net, $2.00

HOBBS, W. R. P. The Arithmetic of Electrical Measurements. With numerous examples, fully worked. *Twelfth Edition.* 12mo., cloth, 126 pp...50 cents

HOMANS, J. E. A B C of the Telephone. With 269 Illustrations. 12mo., cloth, 352 pp..$1.00

HOPKINS, N. M. Experimental Electrochemistry, Theoretically and Practically Treated. Profusely illustrated with 130 new drawings, diagrams, and photographs, accompanied by a Bibliography. Illustrated. 8vo , cloth, 298 pp...................................Net, $3.00

HOUSTON, EDWIN J. A Dictionary of Electrical Words, Terms, and Phrases. *Fourth Edition, rewritten and greatly enlarged.* 582 Illustrations. 4to., cloth................................Net, $7.00

A Pocket Dictionary of Electrical Words, Terms, and Phrases. 12mo., cloth, 950 pp. ....Net, $2.50

HUTCHINSON, R. W., Jr. Long-Distance Electric Power Transmission: Being a Treatise on the Hydro-Electric Generation of Energy; Its Transformation, Transmission, and Distribution. *Second Edition.* Illustrated. 12mo., cloth, 350 pp......................Net, $3.00

HUTCHINSON, R. W., Jr. and IHLSENG, M. C. Electricity in Mining. Being a theoretical and practical treatise on the construction, operation, and maintenance of electrical mining machinery. 12mo., cloth...*In Press*

INCANDESCENT ELECTRIC LIGHTING. A Practical Description of the Edison System, by H. Latimer. To which is added: The Design and Operation of Incandescent Stations, by C. J. Field; A Description of the Edison Electrolyte Meter, by A. E. Kennelly; and a Paper on the Maximum Efficiency of Incandescent Lamps, by T. W. Howell. *Fifth Edition.* Illustrated. 16mo., cloth, 140 pp. (No. 57 Van Nostrand's Science Series.)......................50 cents

INDUCTION COILS: How Made and How Used. *Eleventh Edition.* Illustrated. 16mo., cloth, 123 pp. (No. 53 Van Nostrand's Science Series.)...50 cents

JEHL, FRANCIS. The Manufacture of Carbons for Electric Lighting and other purposes. Illustrated with numerous Diagrams, Tables, and Folding Plates. 8vo., cloth, 232 pp.................Net, $4.00

JONES, HARRY C. The Electrical Nature of Matter and Radioactivity. *Second Edition, revised and enlarged.* 12mo., cloth, 218 pp..$2.00

KAPP, GISBERT. Electrical Transmission of Energy and its Transformation, Subdivision, and Distribution. A Practical Handbook. *Fourth Edition, thoroughly revised.* Illustrated. 12mo., cloth, 445 pp..$3.50

Alternate-Current Machinery. Illustrated. 16mo., cloth, 190 pp. (No. 96 Van Nostrand's Science Series.)......................50 cents

Dynamos, Alternators and Transformers. Illustrated. 8vo., cloth, 507 pp..$4.00

KELSEY, W. R. Continuous-Current Dynamos and Motors, and their Control; being a series of articles reprinted from the "Practical Engineer," and completed by W. R. Kelsey, B.Sc. With Tables, Figures, and Diagrams. 8vo., cloth, 439 pp...............$2.50

KEMPE, H. R. A Handbook of Electrical Testing. *Seventh Edition, revised and enlarged.* Illustrated. 8vo., cloth, 706 pp...Net, $6.00

127

KENNEDY, R. Modern Engines and Power Generators. Illustrated. 8vo., cloth, 5 vols. Each. The set, $15.00................$3.50

Electrical Installations of Electric Light, Power, and Traction Machinery. Illustrated. 8vo., cloth, 5 vols. Each.....................$3.50

KENNELLY, A. E. Theoretical Elements of Electro-Dynamic Machinery. Vol. I. Illustrated. 8vo., cloth, 90 pp....................$1.50

KERSHAW, J. B. C. The Electric Furnace in Iron and Steel Production. Illustrated. 8vo., cloth, 74 pp........................Net, $1.50

Electrometallurgy. Illustrated. 8vo., cloth, 303 pp. (Van Nostrand's Westminster Series.)..........................Net, $2.00

KINZBRUNNER, C. Continuous-Current Armatures; their Winding and Construction. 79 Illustrations. 8vo., cloth, 80 pp......Net, $1.50

Alternate-Current Windings; their Theory and Construction. 89 Illustrations. 8vo., cloth, 80 pp..........................Net, $1.50

KOESTER, F. Hydroelectric Developments and Engineering. A practical and theoretical treatise on the development, design, construction, equipment and operation of hydroelectric transmission plants. 500 illustrations. 4to., cloth, 475 pp......................Net, $5.00

—— Steam-Electric Power Plants. A practical treatise on the design of central light and power stations and their economical construction and operation. Fully Illustrated. 4to., cloth, 455 pp..Net, $5.00

LARNER, E. T. The Principles of Alternating Currents for Students of Electrical Engineering. Illustrated with Diagrams. 12mo., cloth, 144 pp..Net, $1.50

LEMSTROM, S. Electricity in Agriculture and Horticulture. Illustrated. 8vo., cloth...Net, $1.50

LIVERMORE, V. P., and WILLIAMS, J. How to Become a Competent Motorman: Being a practical treatise on the proper method of operating a street-railway motor-car; also giving details how to overcome certain defects. *Second Edition.* Illustrated. 16mo., cloth, 247 pp..Net, $1.00

LOCKWOOD, T. D. Electricity, Magnetism, and Electro-Telegraphy. A Practical Guide and Handbook of General Information for Electrical Students, Operators, and Inspectors. *Fourth Edition.* Illustrated. 8vo., cloth, 374 pp..............................$2.50

128

LODGE, OLIVER J. Signalling Across Space Without Wires: Being a description of the work of Hertz and his successors. *Third Edition.* Illustrated. 8vo., cloth..............................Net, $2.00

LORING, A. E. A Handbook of the Electro-Magnetic Telegraph. *Fourth Edition, revised.* Illustrated. 16mo., cloth, 116 pp. (No. 39 Van Nostrand's Science Series.)......................50 cents

LUPTON, A., PARR, G. D. A., and PERKIN, H. Electricity Applied to Mining. *Second Edition.* With Tables, Diagrams, and Folding Plates. 8vo., cloth, 320 pp.........................Net, $4.50

MAILLOUX, C. O. Electric Traction Machinery. Illustrated. 8vo., cloth..*In Press*

MANSFIELD, A. N. Electromagnets: Their Design and Construction. *Second Edition.* Illustrated. 16mo., cloth, 155 pp. (No. 64 Van Nostrand's Science Series.)...........................50 cents

MASSIE, W. W., and UNDERHILL, C. R. Wireless Telegraphy and Telephony Popularly Explained. With a chapter by Nikola Tesla. Illustrated. 12mo., cloth, 82 pp.....................Net, $1.00

MAURICE, W. Electrical Blasting Apparatus and Explosives, with special reference to colliery practice. Illustrated. 8vo., cloth, 167 pp..Net, $3.50

—— The Shot Firer's Guide. A practical manual on blasting and the prevention of blasting accidents. 78 illustrations. 8vo., cloth, 212 pp..Net, $1.50

MAVER, WM., Jr. American Telegraphy and Encyclopedia of the Telegraph Systems, Apparatus, Operations. *Fifth Edition, revised.* 450 Illustrations. 8vo., cloth, 656 pp.....................Net, $5.00

MONCKTON, C. C. F. Radio Telegraphy. 173 Illustrations. 8vo., cloth, 272 pp. (Van Nostrand's Westminster Series.)....Net, $2.00

MORGAN, ALFRED P. Wireless Telegraph Construction for Amateurs. 153 illustrations. 12mo., cloth, 220 pp.................Net, $1.50

MUNRO, J., and JAMIESON, A. A Pocket-Book of Electrical Rules and Tables for the Use of Electricians, Engineers, and Electrometallurgists. *Eighteenth Revised Edition.* 32mo., leather, 735 pp..........$2.50

129

NIPHER, FRANCIS E. Theory of Magnetic Measurements. With an Appendix on the Method of Least Squares. Illustrated. 12mo., cloth, 94 pp. $1.00

NOLL, AUGUSTUS. How to Wire Buildings. A Manual of the Art of Interior Wiring. *Fourth Edition.* Illustrated. 12mo., cloth, 165 pp. $1.50

OHM, G. S. The Galvanic Circuit Investigated Mathematically. Berlin, 1827. Translated by William Francis. With Preface and Notes by Thos. D. Lockwood. *Second Edition.* Illustrated. 16mo., cloth, 269 pp. (No. 102 Van Nostrand's Science Series.).........50 cents

OLSSON, ANDREW. Motor Control as used in Connection with Turret Turning and Gun Elevating. (The Ward Leonard System.) 13 illustrations. 12mo., paper, 27 pp. (U. S. Navy Electrical Series No. 1.)..Net, .50

OUDIN, MAURICE A. Standard Polyphase Apparatus and Systems. *Fifth Edition, revised.* Illustrated with many Photo-reproductions, Diagrams, and Tables. 8vo., cloth, 369 pp.............Net, $3.00

PALAZ, A. Treatise on Industrial Photometry. Specially applied to Electric Lighting. Translated from the French by G. W. Patterson, Jr., and M. R. Patterson. *Second Edition.* Fully Illustrated. 8vo., cloth, 324 pp. $4.00

PARR, G. D. A. Electrical Engineering Measuring Instruments for Commercial and Laboratory Purposes. With 370 Diagrams and Engravings. 8vo., cloth, 328 pp............................Net, $3.50

PARSHALL, H. F., and HOBART, H. M. Armature Windings of Electric Machines. *Third Edition.* With 140 full-page Plates, 65 Tables, and 165 pages of descriptive letter-press. 4to., cloth, 300 pp. .$7.50

Electric Railway Engineering. With 437 Figures and Diagrams and many Tables. 4to., cloth, 475 pp................Net, $10.00

Electric Machine Design. Being a revised and enlarged edition of "Electric Generators." 648 Illustrations. 4to., half morocco, 601 pp..Net, $12.50

PERRINE, F. A. C. Conductors for Electrical Distribution: Their Manufacture and Materials, the Calculation of Circuits, Pole-Line Construction, Underground Working, and other Uses. *Second Edition.* Illustrated. 8vo., cloth, 287 pp........................Net, $3.50

POOLE, C. P. The Wiring Handbook with Complete Labor-saving Tables and Digest of Underwriters' Rules. Illustrated. 12mo., leather, 85 pp...Net, $1.00

POPE, F. L. Modern Practice of the Electric Telegraph. A Handbook for Electricians and Operators. *Seventeenth Edition.* Illustrated. 8vo., cloth, 234 pp......................................$1.50

RAPHAEL, F. C. Localization of Faults in Electric Light Mains. *Second Edition, revised.* Illustrated. 8vo., cloth, 205 pp......Net, $3.00

RAYMOND, E. B. Alternating-Current Engineering, Practically Treated. *Third Edition, revised.* With many Figures and Diagrams. 8vo., cloth, 244 pp.......................................Net, $2.50

RICHARDSON, S. S. Magnetism and Electricity and the Principles of Electrical Measurement. Illustrated. 12mo., cloth, 596 pp..Net, $2.00

ROBERTS, J. Laboratory Work in Electrical Engineering—Preliminary Grade. A series of laboratory experiments for first- and second-year students in electrical engineering. Illustrated with many Diagrams. 8vo., cloth, 218 pp..................................Net, $2.00

ROLLINS, W. Notes on X-Light. Printed on deckle edge Japan paper. 400 pp. of text, 152 full-page plates. 8vo., cloth.......Net, $7.50

RUHMER, ERNST. Wireless Telephony in Theory and Practice. Translated from the German by James Erskine-Murray. Illustrated. 8vo., cloth, 224 pp...................................Net, $3.50

RUSSELL, A. The Theory of Electric Cables and Networks. 71 Illustrations. 8vo., cloth, 275 pp.........................Net, $3.00

SALOMONS, DAVID. Electric-Light Installations. A Practical Handbook. Illustrated. 12mo., cloth.
Vol. I.: **Management of Accumulators.** *Ninth Edition.* 178 pp.$2.50
Vol. II.: **Apparatus.** *Seventh Edition.* 318 pp...............$2.25
Vol. III.: **Application.** *Seventh Edition.* 234 pp.............$1.50

SCHELLEN, H. Magneto-Electric and Dynamo-Electric Machines. Their Construction and Practical Application to Electric Lighting and the Transmission of Power. Translated from the Third German Edition by N. S. Keith and Percy Neymann. With Additions and Notes relating to American Machines, by N. S. Keith. Vol. I. With 353 Illustrations. *Third Edition.* 8vo., cloth, 518 pp.......$5.00

131

SEVER, G. F. Electrical Engineering Experiments and Tests on Direct-Current Machinery. *Second Edition, enlarged.* With Diagrams and Figures. 8vo., pamphlet, 75 pp......................Net, $1.00

SEVER, G. F., and TOWNSEND, F. Laboratory and Factory Tests in Electrical Engineering. *Second Edition, revised and enlarged.* Illustrated. 8vo., cloth, 269 pp...........................Net, $2.50

SEWALL, C. H. Wireless Telegraphy. With Diagrams and Figures. *Second Edition, corrected.* Illustrated. 8vo., cloth, 229 pp. . Net, $2.00

Lessons in Telegraphy. Illustrated. 12mo., cloth, 104 pp. . Net, $1.00

SEWELL, T. Elements of Electrical Engineering. *Third Edition, revised.* Illustrated. 8vo., cloth, 444 pp...................$3.00

The Construction of Dynamos (Alternating and Direct Current). A Text-book for students, engineering contractors, and electricians-in-charge. Illustrated. 8vo., cloth, 316 pp.................:..$3.00

SHAW, P. E. A First-Year Course of Practical Magnetism and Electricity. Specially adapted to the wants of technical students. Illustrated. 8vo., cloth, 66 pp. interleaved for note taking...........Net, $1.00

SHELDON, S., and HAUSMANN, E. Dynamo-Electric Machinery: Its Construction, Design, and Operation.
Vol. I.: Direct-Current Machines. *Eighth Edition, completely re-written.* Illustrated. 8vo., cloth, 310 pp.....................Net, $2.50

SHELDON, S., MASON, H., and HAUSMANN, E. Alternating-Current Machines: Being the second volume of "Dynamo-Electric Machinery; its Construction, Design, and Operation." With many Diagrams and Figures. (Binding uniform with Volume I.) *Seventh Edition, rewritten.* 8vo., cloth, 353 pp........Net, $2.50

SLOANE, T. O'CONOR. Standard Electrical Dictionary. 300 Illustrations. 12mo., cloth, 682 pp.............................$3.00

——— Elementary Electrical Calculations. A Manual of Simple Engineering Mathematics, covering the whole field of Direct Current Calculations, the basis of Alternating Current Mathematics, Networks, and typical cases of Circuits, with Appendices on special subjects. 8vo., cloth. Illustrated. 304 pp...........Net, $2.00

SNELL, ALBION T. Electric Motive Power. The Transmission and Distribution of Electric Power by Continuous and Alternating Currents. With a Section on the Applications of Electricity to Mining Work. *Second Edition.* Illustrated. 8vo., cloth, 411 pp.......Net, $4.00

132

SODDY, F. Radio-Activity; an Elementary Treatise from the Standpoint of the Disintegration Theory. Fully Illustrated. 8vo., cloth, 214 pp..Net, $3.00

SOLOMON, MAURICE. Electric Lamps. Illustrated. 8vo., cloth. (Van Nostrand's Westminster Series.)....................Net, $2.00

STEWART, A. Modern Polyphase Machinery. Illustrated. 12mo., cloth, 296 pp.....................................Net, $2.00

SWINBURNE, JAS., and WORDINGHAM, C. H. The Measurement of Electric Currents. Electrical Measuring Instruments. Meters for Electrical Energy. Edited, with Preface, by T. Commerford Martin. Folding Plate and Numerous Illustrations. 16mo., cloth, 241 pp. (No. 109 Van Nostrand's Science Series.)...............50 cents

SWOOPE, C. WALTON. Lessons in Practical Electricity: Principles, Experiments, and Arithmetical Problems. An Elementary Textbook. With numerous Tables, Formulæ, and two large Instruction Plates. *Eleventh Edition, revised.* Illustrated. 8vo., cloth, 462 pp.
Net, $2.00

THOM, C., and JONES, W. H. Telegraphic Connections, embracing recent methods in Quadruplex Telegraphy. 20 Colored Plates. 8vo., cloth, 59 pp..$1.50

THOMPSON, S. P. Dynamo-Electric Machinery. With an Introduction and Notes by Frank L. Pope and H. R. Butler. Fully Illustrated. 16mo., cloth, 214 pp. (No. 66 Van Nostrand's Science Series.)
50 cents

Recent Progress in Dynamo-Electric Machines. Being a Supplement to "Dynamo-Electric Machinery." Illustrated. 16mo., cloth, 113 pp. (No. 75 Van Nostrand's Science Series.)..................50 cents

TOWNSEND, FITZHUGH. Alternating Current Engineering. Illustrated. 8vo., paper, 32 pp.......................Net, 75 cents

UNDERHILL, C. R. Solenoids, Electromagnets and Electromagnetic Windings. 218 Illustrations. 12mo., cloth, 345 pp.....Net, $2.00

URQUHART, J. W. Dynamo Construction. A Practical Handbook for the use of Engineer Constructors and Electricians in Charge. Illustrated. 12mo., cloth....................................$3.00

Electric Ship-Lighting. A Handbook on the Practical Fitting and Running of Ship's Electrical Plant, for the use of Ship Owners and Builders, Marine Electricians, and Sea-going Engineers in Charge. 88 Illustrations. 12mo., cloth, 308 pp......................$3.00

Electric-Light Fitting. A Handbook for Working Electrical Engineers, embodying Practical Notes on Installation Management. *Second Edition*, with additional chapters. With numerous Illustrations. 12mo., cloth..$2.00

Electroplating. A Practical Handbook. *Fifth Edition.* Illustrated. 12mo., cloth, 230 pp....................................$2.00

Electrotyping. Illustrated. 12mo., cloth, 228 pp.............$2.00

WADE, E. J. Secondary Batteries: Their Theory, Construction, and Use. *Second Edition, corrected.* 265 Illustrations. 8vo., cloth, 501 pp.
Net, $4.00

WADSWORTH, C. Electric Battery Ignition. 15 Illustrations. 16mo. paper...*In Press*

WALKER, FREDERICK. Practical Dynamo-Building for Amateurs. How to Wind for any Output. *Third Edition.* Illustrated. 16mo., cloth, 104 pp. (No. 98 Van Nostrand's Science Series.).....50 cents

WALKER, SYDNEY F. Electricity in Homes and Workshops. A Practical Treatise on Auxiliary Electrical Apparatus. *Fourth Edition.* Illustrated. 12mo., cloth, 358 pp........................$2.00

Electricity in Mining. Illustrated. 8vo., cloth, 385 pp........$3.50

WALLING, B. T., and MARTIN, JULIUS. Electrical Installations of the United States Navy. With many Diagrams and Engravings. 8vo., cloth, 648 pp...$6.00

WALMSLEY, R. M. Electricity in the Service of Man. A Popular and Practical Treatise on the Application of Electricity in Modern Life. Illustrated. 8vo., cloth, 1208 pp.....................Net, $4.50

WATT, ALEXANDER. Electroplating and Refining of Metals. *New Edition*, rewritten by Arnold Philip. Illustrated. 8vo., cloth, 677 pp...Net, $4.50

Electrometallurgy. *Fifteenth Edition.* Illustrated. 12mo., cloth, 225 pp...$1.00

134

WEBB, H. L. A Practical Guide to the Testing of Insulated Wires and Cables. *Fifth Edition.* Illustrated. 12mo., cloth, 118 pp.....$1.00

WEEKS, R. W. The Design of Alternate-Current Transformer.
New Edition in Press

WEYMOUTH, F. MARTEN. Drum Armatures and Commutators. (Theory and Practice.) A complete treatise on the theory and construction of drum-winding, and of commutators for closed-coil armatures, together with a full résumé of some of the principal points involved in their design, and an exposition of armature reactions and sparking. Illustrated. 8vo., cloth, 295 pp.........Net, $3.00

WILKINSON, H. D. Submarine Cable Laying, Repairing and Testing. *Second Edition, completely revised.* 313 Illustrations. 8vo., cloth, 580 pp..Net, $6.00

YOUNG, J. ELTON. Electrical Testing for Telegraph Engineers. Illustrated. 8vo., cloth, 264 pp......................Net, $4.00

ZEIDLER, J., and LUSTGARTEN, J. Electric Arc Lamps: Their Principles, Construction and Working. 160 Illustrations. 8vo., cloth, 188 pp...Net, $2.00

A **96**-page Catalog of Books on Electricity, classified by subjects, will be furnished gratis, postage prepaid, on application.

NOTES

www.ingramcontent.com/pod-product-compliance
Lightning Source LLC
Chambersburg PA
CBHW051527170526
45165CB00002B/636